YOUR FACE BELONGS TO US

YOUR FACE BELONGS TO US

THE SECRETIVE STARTUP DISMANTLING YOUR PRIVACY

KASHMIR HILL

SIMON & SCHUSTER

London · New York · Sydney · Toronto · New Delhi

First published in the United States by Random House, an imprint and division of
Penguin Random House LLC, New York, 2023
First published in Great Britain by Simon & Schuster UK Ltd, 2023

1 3 5 7 9 10 8 6 4 2

Simon & Schuster UK Ltd
1st Floor
222 Gray's Inn Road
London WC1X 8HB

www.simonandschuster.co.uk
www.simonandschuster.com.au
www.simonandschuster.co.in

Simon & Schuster Australia, Sydney
Simon & Schuster India, New Delhi

A CIP catalogue record for this book
is available from the British Library

Hardback ISBN: 978-1-3985-0917-7
Trade Paperback ISBN: 978-1-3985-0918-4
eBook ISBN: 978-1-3985-0919-1

Book design by Fritz Metsch

Printed and Bound in the UK using 100%
Renewable Electricity at CPI Group (UK) Ltd

CONTENTS

Prologue: The Tip vii

PART I: THE FACE RACE

1 : A Strange Kind of Love 3

2 : The Roots (350 B.C.–1880s) 17

3 : "Fatface Is Real" 27

4 : If At First You Don't Succeed (1956–1991) 36

5 : A Disturbing Proposal 50

6 : The Snooper Bowl (2001) 60

7 : The Supercomputer Under the Bed 72

8 : The Only Guy Who Saw It Coming (2006–2008) 82

9 : Death to Smartcheckr 88

PART II: TECHNICAL SWEETNESS

10 : The Line Google Wouldn't Cross (2009–2011) 99

11 : Finding Mr. Right 111

12 : The Watchdog Barks (2011–2012) 121

13 : Going Viral 128

14 : "You Know What's Really Creepy?" (2011–2019) 140

15 : Caught in a Dragnet 153

16 : Read All About It 160

PART III: FUTURE SHOCK

17 : "Why the Fuck Am I Here?" (2020) 169

18 : A Different Reason to Wear a Mask 185

19 : I Have a Complaint 190

20 : The Darkest Impulses 197

21 : Code Red (or, Floyd Abrams v. the ACLU) 202

22 : The Future Is Unevenly Distributed 214

23 : A Rickety Surveillance State 228

24 : Fighting Back 237

25 : Tech Issues 245

Acknowledgments 253

A Note on Sources 255

Notes 257

Index 315

PROLOGUE

THE TIP

IN NOVEMBER 2019, I had just become a reporter at *The New York Times* when I got a tip that seemed too outrageous to be true: A mysterious company called Clearview AI claimed it could identify just about anyone based only on a snapshot of their face.

I was in a hotel room in Switzerland, six months pregnant, when I got the email. It was the end of a long day and I was tired but the email gave me a jolt. My source had unearthed a legal memo marked "Privileged & Confidential" in which a lawyer for Clearview had said that the company had scraped billions of photos from the public web, including social media sites such as Facebook, Instagram, and LinkedIn, to create a revolutionary app. Give Clearview a photo of a random person on the street, and it would spit back all the places on the internet where it had spotted their face, potentially revealing not just their name but other personal details about their life. The company was selling this superpower to police departments around the country but trying to keep its existence a secret.

Not so long ago, automated facial recognition was a dystopian technology that most people associated only with science fiction novels or movies such as *Minority Report*. Engineers first sought to make it a reality in the 1960s, attempting to program an early computer to match someone's portrait to a larger database of people's faces. In the early 2000s, police began experimenting with it to search mug shot databases for the faces of unknown criminal suspects. But the technology had largely proved disappointing. Its performance varied across race, gender, and age, and even state-of-the-art algorithms struggled to do something as simple as matching a mug shot to a grainy ATM surveillance still. Clearview claimed to be different, touting a "98.6 percent accuracy rate" and an enormous collection of photos unlike anything the police had used before.

This is huge if true, I thought as I read and reread the Clearview memo that had never been meant to be public. I had been covering privacy, and its steady erosion, for more than a decade. I often describe my beat as "the looming tech dystopia—and how we can try to avoid it," but I'd never seen such an audacious attack on anonymity before.

Privacy, a word that is notoriously hard to define, was most famously described in a *Harvard Law Review* article in 1890 as "the right to be let alone." The two lawyers who wrote the article, Samuel D. Warren, Jr., and Louis D. Brandeis, called for the right to privacy to be protected by law, along with those other rights—to life, liberty, and private property—that had already been enshrined. They were inspired by a then-novel technology—the portable Eastman Kodak film camera, invented in 1888, which made it possible to take a camera outside a studio for "instant" photos of daily life—as well as by people like me, a meddlesome member of the press.

"Instantaneous photographs and newspaper enterprise have invaded the sacred precincts of private and domestic life," wrote Warren and Brandeis, "and numerous mechanical devices threaten to make good the prediction that 'what is whispered in the closet shall be proclaimed from the house-tops.' "

This article is among the most famous legal essays ever written, and Louis Brandeis went on to join the Supreme Court. Yet privacy never got the kind of protection Warren and Brandeis said that it deserved. More than a century later, there is still no overarching law guaranteeing Americans control over what photos are taken of them, what is written about them, or what is done with their personal data. Meanwhile, companies based in the United States—and other countries with weak privacy laws—are creating ever more powerful and invasive technologies.

Facial recognition had been on my radar for a while. Throughout my career, at places such as *Forbes* and *Gizmodo,* I had covered major new offerings from billion-dollar companies: Facebook automatically tagging your friends in photos; Apple and Google letting people look at their phones to unlock them; digital billboards from Microsoft and Intel with cameras that detected age and gender to show passersby appropriate ads.

I had written about the way this sometimes clunky and error-prone technology excited law enforcement and industry but terrified privacy-conscious citizens. As I digested what Clearview claimed it could do, I thought back to a federal workshop I'd attended years earlier in Washington, D.C., where industry representatives, government officials, and privacy advocates had sat down to hammer out the rules of the road. The one thing they all agreed on was that *no one* should roll out an application to identify strangers. It was too dangerous, they said. A weirdo at a bar could snap your photo and within seconds know who your friends were and where you lived. It could be used to identify antigovernment protesters or women who walked into Planned Parenthood clinics. It would be a weapon for harassment and intimidation. Accurate facial recognition, on the scale of hundreds of millions or billions of people, was the third rail of the technology. And now Clearview, an unknown player in the field, claimed to have built it.

I was skeptical. Startups are notorious for making grandiose claims that turn out to be snake oil. Even Steve Jobs famously faked the capabilities of the original iPhone when he first revealed it onstage in 2007.* We tend to believe that computers have almost magical powers, that they can figure out the solution to any problem and, with enough data, eventually solve it better than humans can. So investors, customers, and the public can be tricked by outrageous claims and some digital sleight of hand by companies that aspire to do something great but aren't quite there yet.

But in this confidential legal memo, Clearview's high-profile lawyer, Paul Clement, who had been the solicitor general of the United States under President George W. Bush, claimed to have tried out the product with attorneys at his firm and "found that it returns fast and

* Steve Jobs pulled a fast one, hiding the prototype iPhone's memory problems and frequent crashes by having his engineers spend countless hours on finding a "golden path"—a specific sequence of tasks the phone could do without glitching.

accurate search results." Clement wrote that more than two hundred law enforcement agencies were already using the tool and that he'd determined that they "do not violate the federal Constitution or relevant existing state biometric and privacy laws when using Clearview for its intended purpose." Not only were hundreds of police departments using this tech in secret, but the company had hired a fancy lawyer to reassure officers that they weren't committing a crime by doing so.

I returned to New York with an impending birth as a deadline. I had three months to get to the bottom of this story, and the deeper I dug, the stranger it got.

THE COMPANY'S ONLINE presence was limited to a simple blue website with a Pac-Man-esque logo—the *C* chomping down on the *V*—and the tagline "Artificial Intelligence for a better world." There wasn't much else there, just a form to "request access" (which I filled out and sent to no avail) and an address in New York City.

A search of LinkedIn, the professional networking site where tech employees can be relied on to brag about their jobs, came up empty save for a single person named John Good. Though he looked middle-aged in his profile photo, he had only one job on his résumé: "sales manager at Clearview AI." Most professionals on LinkedIn are connected to hundreds of people; this guy was connected to two. Generic name. Skimpy résumé. Almost no network. Was this even a real person?

I sent Mr. Good a message but never got a response.

So I decided to go door knocking. I mapped the address from the company's website and discovered that it was in midtown Manhattan, just two blocks away from the *New York Times* building. On a cold, gray afternoon, I walked there, slowly, because I was in the stage of the pregnancy where walking too fast gave me Braxton-Hicks contractions.

When I arrived at the point on the sidewalk where Google Maps directed me, the mystery deepened. The building where Clearview was supposedly headquartered did not exist.

The company's listed address was 145 West Forty-first Street. There was a delivery dock at 143 West Forty-first Street and next to it on the corner of Broadway, where 145 should have been, an outpost of the co-working giant WeWork—but its address was 1460 Broadway. Thinking that Clearview must be renting a WeWork office, I popped my head in and asked the receptionist, who said there was no such company there.

It was like something out of Harry Potter. Was there a magic door I was missing?

I reached out to Clearview's lawyer, Paul Clement, to see if he'd actually written the legal memo for this company with one fake employee working out of a nonexistent building. Despite repeated calls and emails, I got no response.

I did some digging—searching for mention of Clearview on government websites and those that track startup investments—and found a few other people with links to the company. A brief listing on the venture capital–tracking site PitchBook claimed that the company had two investors—one I had never heard of before and another who was shockingly recognizable: The polarizing billionaire Peter Thiel, who had co-founded the payments platform PayPal, backed the social network Facebook early on, and created the data-mining juggernaut Palantir. Thiel had a Midas touch when it came to tech investments, but he was the subject of immense criticism for his contrarian views, his support of Donald Trump for the presidency, and his funding of a legal campaign against the news blog Gawker that had ultimately put the media outlet out of business.

Thiel, like everyone else I reached out to, gave me the cold shoulder.

BUSINESS FILINGS REVEALED that the company had incorporated in Delaware in 2018 using an address on the Upper West Side. I bundled up and headed to the subway. The train, a local on the C line, was not crowded, and my anticipation grew with each stop. When I got off, a little past the Natural History Museum, I discovered that the address was on an unusually quiet street, next to the castlelike Dakota Apartments building. Eager to knock on an actual door that might unravel the mystery, I couldn't help but speed my steps, which made the muscles across my belly clench in protest.

The building had an Art Deco exterior, with a revolving glass door, and I could see individual balconies on the floors above that gave a distinct residential impression. The lounge looked homey with a surprising number of couches and a Christmas tree. A uniformed doorman greeted me at the entrance and asked whom I was there to see.

"Clearview AI in Suite 23-S," I said.

He looked at me quizzically. "There are no businesses here," he said. "That's someone's home."

He wouldn't let me up. Yet another dead end.

THEN ONE DAY I logged in to Facebook to discover a message from a "friend" named Keith. I didn't remember him, but he mentioned that we had met a decade earlier at a gala for Italian Americans. Back then, I'd been more cavalier about my privacy and said yes to just about any "friend request."

"I understand you are looking to connect with Clearview," Keith wrote. "I know the company, they are great. How can I be of help?"

Keith worked at a real estate company in New York and had no obvious ties to the facial recognition startup. I had many questions—foremost among them being how he knew I was looking into the company and whether he knew the identity of the technological mastermind who had supposedly built the futuristic app—so I asked him for his phone number.

He didn't respond.

I asked again two days later.

Silence.

As it became clear that the company wasn't going to talk to me, I tried a new approach: Find out whether the tool was as effective as advertised. I tracked down police officers who had used it, starting with a detective sergeant in Gainesville, Florida, named Nicholas Ferrara. For the first time, I found someone who wanted to discuss Clearview with me. In fact, he could not have been more excited about it.

"I love it. It's amazing," he said. "I'd be their spokesperson if they wanted me."

Ferrara had first heard about the company when it had advertised on CrimeDex, a listserv for investigators who specialize in financial crimes. The company described itself as "Google search for faces" and offered a free thirty-day trial. All he had to do was send an email to sign up.

When he got access to Clearview, he took a selfie and ran a search on it. The first hit was his Venmo profile photo, along with a link to his page on the mobile payments site. He was impressed.

Ferrara had dozens of unsolved cases in which his only lead was a photo of a fraudster standing at an ATM or bank counter. He had previously run the photos through a state-provided face recognition tool but come up empty. Now he uploaded them to Clearview. He got match after match after match. He identified thirty suspects, just like that. It was incredible.

The government systems he had previously used required a person's full face, ideally gazing directly at the camera, but Clearview worked on photos of suspects wearing hats or glasses, even when only part of their face was visible. The other systems could comb only through the faces of people who had been arrested or who had gotten their driver's licenses in Florida, but Clearview found people who had no ties to the state and even people from other countries. He was blown away.

One weekend night, while Ferrara was patrolling in downtown Gainesville, he ran into a group of college students hanging around outside a bar. He struck up a conversation about how their night was going and then asked them if he could do an experiment: Could he try to identify them?

The students were game, so, in the dim street light, he snapped their photos one by one. Clearview served up their Facebook and Instagram pages, and he identified four out of five of the students by name. The astounded students thought it was an awesome party trick. Can anyone do that? they asked. Ferrara explained that it was only available to law enforcement.

"That's very Big Brother," one replied.

I wanted to see the results from this amazing app myself. Ferrara volunteered to do a demo, saying I just needed to send him a photo. It felt odd to email my photos to a police officer I'd never actually met in person, but I am almost always willing to sacrifice my own privacy for a story. I sent three: one where I was wearing sunglasses and a hat, one with my eyes closed, and a third of me smiling.

Then I waited for a response.

It never came.

I got in touch with another investigator, Captain Daniel Zientek, who worked in The Woodlands, Texas. He had recently used Clearview to identify an alleged rapist, based on a photo the victim had taken earlier in the evening. He said it had worked for him pretty much every time he'd run a search. The only time it didn't, he said, was for online ghosts. "If you don't have a photo on the internet, it's not going to find you," Zientek said.

Anticipating the controversy over the idea of fingering someone for a crime based solely on a photo, Zientek told me emphatically that he would never arrest someone based only on a Clearview identification. "You still have to build a case," he said.

He offered to run my photo to show me how it worked. I emailed one to him, and he wrote back immediately saying it had no matches.

That was odd. There were tons of photos of me online.

I sent him a different photo, and again it came back with no matches.

Zientek was surprised. So was I. I thought this was supposed to be "game changing" facial recognition. It made no sense. He suggested that it was a technical issue; maybe the company's servers were down.

Then he, too, stopped replying to me.

I REDOUBLED MY efforts to get someone connected to Clearview to talk to me. I doorstepped Kirenaga Partners, the investor listed alongside Peter Thiel on PitchBook. It was a tiny venture capital firm based forty minutes outside New York City. On a rainy Tuesday morning, I took the Metro North train from Manhattan to a wealthy suburb called Bronxville, where I walked a few blocks to a two-story office building on a quaint retail block across from a hospital. I climbed the stairs to the second floor and found the Kirenaga Partners logo—a samurai sword wrapped in silk—at the end of a long, quiet hallway lined with office doors. There was no response when I knocked and no ringing inside when I called the office's number.

A neighbor and a deliveryman told me that the office was rarely occupied. After lingering awkwardly in the hallway for almost an hour, I decided to give up and head back to Manhattan, but as I was descending the staircase, two men walked in, one in a lilac shirt and a dark suit and another in a gray-and-pink ensemble. They had the look of money about them, and when they made eye contact with me, I asked if they were with Kirenaga Partners.

"Yes," said the dark-haired one in lilac, with a surprised smile on his face. "Who are you?"

When I introduced myself, his smile faded. "Oh, Clearview's lawyers said we're not supposed to talk to you."

"I came all this way," I pleaded, making sure my pregnant belly was visible beneath my red winter coat. They looked at each other, and then the one in lilac—who turned out to be the firm's founder, David Scalzo—seemingly trapped by his own sense of decency, offered me water and begrudgingly agreed to talk briefly and off the record. We went upstairs and they let me into the cold office, where college test prep materials sat on the receptionist desk, and into a small conference room. Scalzo brought me a bottle of water and sat down across from me.

We talked for a few minutes before I dared start taking notes. Despite his initial reluctance, Scalzo was soon enthusing about his promising investment. After I revealed that I'd spoken with officers who had used the tool and that I would be writing a story with or without the cooperation of the company, he agreed to go on the record to sing the company's praises. He said that the company had not been responding to my requests because it was in "stealth mode" and because one of the founders had some "Gawker history" which he wasn't keen to have unearthed.

He told me that law enforcement loved Clearview but that the company ultimately aimed to offer the face recognition app to *everyone*. I expressed my trepidation about the likelihood that that would bring an end to the possibility of anonymity. "I've come to the conclusion that because information constantly increases, there's never going to be privacy," Scalzo mused. "You can't ban technology. Sure, that might lead to a dystopian future or something, but you can't ban it."

WHEN I ASKED privacy and legal experts about what Clearview AI appeared to have built, they were universally horrified and said that the weaponization possibilities were "endless." A rogue police officer could use Clearview to stalk potential romantic partners. A foreign government could dig up secrets about its opponents to blackmail them or throw them in jail. People at protests could be identified by police, including in countries such as China and Russia with track records of repressing, and even killing, dissidents. With increasingly ubiquitous surveillance cameras, an authoritarian leader could use this power to track rivals' faces, watching where they went and with whom they associated, to find compromising moments or build a dossier to be used against them. Keeping secrets for any reason, even safety, could prove impossible.

Beyond Big Brother and the danger of powerful governments was the even more insidious Little Brother: your neighbor, a possessive partner, a stalker, a stranger who bears you ill will. If it became widely available, Clearview could create a culture of justified paranoia.

It would be as if we were all celebrities with famous faces. We could no longer trust that when we bought condoms, pregnancy tests, or hemorrhoid cream with cash at the pharmacy, we did so anonymously. A sensitive conversation over dinner in a restaurant could be tied back to us by a person seated nearby. Gossip casually relayed over the phone while walking down the street could get blasted out by a stranger on Twitter with your name as the source. A person you thoughtlessly offended in a store or cut off in traffic could snap a photo of your face, find out your name, and write horrible things about you online in an attempt to destroy your reputation. Businesses could discriminate against you in new ways, refusing you entry because of who you worked for or something you once said online.

That an unknown startup heralded this new reality, rather than a technology giant such as Google or Facebook, did not surprise the experts with whom I spoke. "These small companies are able to, under the radar, achieve terrible things," one said. Once they do, it's hard to turn back.

People often defend a new, intimidating technology by saying it is like a knife, simply a tool that can be used for good or evil. But most technologies, from social media to facial recognition, are not neutral; and they are far more complicated than a knife. The countless decisions a creator makes about the architecture of a technology's platform shape the way that users will interact with it. OkCupid encouraged daters to consider people's personalities, for instance, by offering lengthy profiles and quizzes to determine compatibility, whereas Tinder's focus was on appearance, presenting photos to users and telling them to swipe left or right. The people who control a technology that becomes widely used wield great power over our society. Who were the people behind Clearview?

WITH THE HELP of a colleague, I recruited a detective based in Texas who was willing to assist with the investigation, as long as I didn't reveal his name. He went to Clearview's website and requested access.

Unlike me, he got a response within half an hour with instructions on how to create an account for his free trial. All he needed was a police department email address. He ran a few photos of criminal suspects whose identities he already knew, and Clearview nailed them all, linking to photos of the correct people on the web.

He ran his own image through the app. He had purposefully kept photos of himself off the internet for years, so he was shocked when he got a hit: a photo of him in uniform, his face tiny and out of focus. It was cropped from a larger photo, for which there was a link that took him to Twitter. A year earlier, someone had tweeted a photo from a Pride Festival. The Texas investigator had been on patrol at the event, and he appeared in the background of someone else's photo. When he zoomed in, his name badge was legible. He was shocked that a face-searching algorithm this powerful existed. He could see that it would be a godsend to some types of law enforcement, but it had horrible implications for undercover officers if the technology became publicly available.

I told him that I hadn't been able to get a demo yet and that another officer had run my photo and gotten no results. He ran my photo again and confirmed that there were no matches.

Minutes later, his phone rang. It was a number he didn't recognize, with a Virginia area code. He picked up.

"Hello. This is Marko with Clearview AI tech support," said the voice on the other end of the call. "We have some questions. Why are you uploading a *New York Times* reporter's photo?"

"I did?" the officer responded cagily.

"Yes, you ran this Kashmir Hill lady from *The New York Times,*" said Marko. "Do you know her?"

"I'm in Texas," the officer replied. "How would I know her?"

The company representative said it was "a policy violation" to run photos of reporters in the app and deactivated the officer's account. The officer was taken aback, creeped out that his use of the app was that closely monitored. He called immediately to tell me what had happened.

A chill ran through me. It was a shocking demonstration of just how much power this mysterious company wielded. They could not only see who law enforcement was looking for, they could block the results. They controlled who could be found. I suddenly understood why the earlier detectives had gone cold on me.

Though Clearview was doing everything in its power to stay hidden, it was using its technology to spy on me. What else was it capable of?

Concerns about facial recognition had been building for decades. And now the nebulous bogeyman had finally found its form: a small company with mysterious founders and an unfathomably large database. And none of the millions of people who made up that database had given their consent. Clearview AI represents our worst fears, but it also offers, at long last, the opportunity to confront them.

PART I

THE FACE RACE

CHAPTER 1

A STRANGE KIND OF LOVE

In the summer of 2016, in an arena in downtown Cleveland, Ohio, Donald Trump, a real estate mogul who had become famous hosting a reality TV show called *The Apprentice,* was being anointed as the Republican Party's presidential candidate. Trump had no government experience, no filter, and no tolerance for political correctness, but his famous name, right-wing populist views, and uncanny ability to exploit his opponents' weaknesses made him a powerful candidate. It was going to be the strangest political convention in American history, and Hoan Ton-That did not want to miss it.*

Ton-That, twenty-eight, did not look, at first glance, like a Trump supporter. Half Vietnamese and half Australian, he was tall—six feet, one inch—with long, silky black hair and androgynous good looks. He dressed flamboyantly in paisley print shirts and suits in a rainbow of colors made bespoke in Vietnam, where, his father told him, his ancestors had once been royalty.

Ton-That had always followed his curiosity no matter where it had taken him. As a kid, growing up in Melbourne, Australia, he had tried to figure out electricity, plugging an extension cord into itself in the hope that the current would race around in a circle inside. Once, after seeing someone steal a woman's purse, he walked up to the thief and asked why he'd done it. The guy pushed him over and ran away. When Ton-That's dad brought a computer home in the early 1990s, Ton-That became obsessed with it, wanting to take it apart, put it back together, play games on it, and type commands into it. Because his growth

* It's pronounced Hwan Tawn-Tat, much like "Juan."

spurt came late, Ton-That was "little Hoan" to his classmates, and bullies targeted him, particularly those who didn't like the growing Asian demographic in Australia. Ton-That had close friends with whom he played guitar, soccer, and cricket, but he felt like an outsider at school.

When he was fourteen, he taught himself to code, relying on free teaching materials and video lectures that MIT posted online. Sometimes he would skip school, tape a "Do Not Disturb" sign on his bedroom door, and spend the day with virtual programming professors. His mother was perplexed by his love of computers, pushing him to pursue a musical career instead after he placed second in a national competition for his guitar playing. "Very often, you'd find him on the computer and the guitar at the same time," his father, Quynh-Du Ton-That, said. "Back and forth between the keyboard and the guitar."

When it came time for college in Canberra, Ton-That chose a computer engineering program, but he was bored out of his mind by professors teaching a mainstream programming language called Java. Ton-That was into "computer snob languages," such as Lisp and Haskell, but those were about as likely to get him a job as studying Latin or ancient Greek. Instead of doing schoolwork, he created a dating app for fellow students called Touchfelt and spent a lot of time on a new online message forum, eventually called Hacker News, which catered to a subculture of people obsessed with technology and startups. When a California investor named Naval Ravikant put up a post there saying he was investing in social media startups, Ton-That reached out to him.

He told Ravikant about his entrepreneurial ambitions and the lack of appetite for such endeavors in Australia. Ravikant told him he should come to San Francisco. So at just nineteen years old, Ton-That moved there, drawn halfway around the world by the siren song of Silicon Valley.

Ravikant picked up the jet-lagged Ton-That at the airport, found him a friend's couch to sleep on, and talked his ear off about the booming Facebook economy. The social network had just opened its platform to outside developers. For the first time, anyone could build an app for Facebook's then 20 million users, and there was a plethora of vacuous offerings. The epitome of those would be FarmVille, a game that let Facebook users harvest virtual crops and raise digital livestock. When the company behind FarmVille went public, it revealed an annual profit of $91 million on $600 million in revenue, thanks to ads and people paying real money to buy digital cows.

"It's the craziest thing ever," Ravikant told Ton-That. "These apps have forty thousand new users a day."

Ton-That couch surfed for at least three months because he didn't have the credit history needed to rent an apartment, but the San Francisco Bay area, with its eclectic collection of entrepreneurs, musicians, and artists, was otherwise a good fit for him. He loved being in the heart of the tech world, meeting either startup founders or those who worked for them. He became friends with early employees at Twitter, Square, Airbnb, and Uber. "You see a lot of this stuff coming out, and it just gives you a lot of energy," he said.

America's diversity was novel to him. "There wasn't African American culture or Mexican culture in Australia," he said. "I had never heard of a burrito before." And he was shocked, at first, by San Francisco's openly gay and gender-bending culture.

Eventually, though, Ton-That let his hair grow long and embraced a gender-fluid identity himself, though he still preferred he/him pronouns. Before he turned twenty-one, he got married, to a Black woman of Puerto Rican descent with whom he performed in a band. (They were in love, he said, but it was also his path to a green card.)

Desperate to make money and continue funding his stay in the United States, he jettisoned his ideals and got to work cranking out Facebook quiz apps. Rather than a "snob" computer language, he adopted what everyone else was using: a basic, utilitarian language called PHP. Speed was more important than style when it came to cashing in on the latest consumer tech addiction. After one of Ton-That's apps got some traction, Ravikant gave him $50,000 in seed money to keep going.

More than 6 million Facebook users installed Ton-That's banal creations—"Have You Ever," "Would You Rather," "Friend Quiz," and "Romantic Gifts"—which he monetized with little banner ads. Initially, Ton-That could get a Facebook user to invite all of their friends to take the quiz with just a few clicks. But the social network was starting to feel spammy. Every time users logged in, they were bombarded with notifications about their friends' FarmVille activities and every quiz any of their friends had taken. Facebook decided to pull back on third-party developers' ability to send notifications to a user's friends, putting an end to the free marketing that Ton-That and others had come to rely on and drying up their steady stream of new users.

Still, Ton-That couldn't believe how much data Facebook had given him along the way: The people who had installed his apps had ceded their names, their photos, their interests, and their likes, and all the same information for all of their friends. When Facebook opened its platform to developers like Ton-That, it didn't just let new third-party apps in; it let users' data flow out, a sin outsiders fully grasped only a decade later during the Cambridge Analytica scandal. But Ton-That didn't have a plan for all that data, at least not yet.

Soon the Facebook app craze faded, and a new tech addiction arrived in the form of mobile apps. Soon after Ton-That arrived in California, Apple released the iPhone. "It was the first thing I got in San Francisco before I even got an apartment," he said. The iPhone almost immediately became an expensive, must-have gadget, and the App Store became a thriving marketplace, with monied iPhone users willing to hand over their credit card numbers for apps that would make the most of their precious devices. As the smartphone economy began to boom, Ton-That sold his Facebook quiz company, paid Ravikant back, and turned to making iPhone games.

It was fun at first—and lucrative. He created a Pavlovian game called Expando that involved repeatedly tapping the iPhone's screen to blow up digital smiley face balloons, while tilting the phone this way and that to roll the balloons away from orange particles that would pop them. People paid up to $1.99 each to download it, which added up quickly when the game proved popular. But over time, the iPhone game space got more competitive and the amount Ton-That could charge dropped until eventually he had to make the game free and rely on the money he made from annoying banner ads.

At the beginning of the twenty-first century, companies like Google and Facebook convinced much of the planet that the tech industry had a noble plan to connect and enrich the global population, unlock the whole of human knowledge, and make the world a better place. But on the ground in San Francisco, many developers like Ton-That were just trying to strike it rich any dumb way they could.

One Tuesday in February 2009, Ton-That released ViddyHo, a YouTube imitator that required visitors wanting to watch a video to log in using their credentials for Google's chat service. But when they did, ViddyHo used the credentials they provided to send messages *as them* to all their friends, saying, "Hey, check out this video," with a shortened link to ViddyHo that would start the cycle anew. This was similar to a growth hacking trick pioneered by Facebook and LinkedIn—who rummaged through the contact books handed over by their users to email people who hadn't joined yet— but ViddyHo's technique was far more invasive, potentially illegal, and its vaguely porny name didn't help matters. The super-spreader event soon had "ViddyHo virus" trending on Twitter.

"Fast-Spreading Phishing Scam Hits Gmail Users," read a headline on *The New York Times*' Bits blog. The story quoted a security expert who said, "These criminals really know how to get people's attention." Ton-That was taking a shower when his phone started going crazy, friends texting him to urge him to shut ViddyHo down. He did, but it was too late. Google blocked him, his internet service provider dropped his account, internet browsers threw up a warning message to anyone who visited ViddyHo.com, and the tech gossip blog Valleywag got on the case.

Valleywag was part of the Gawker Media empire, an influential collection of snarky news blogs with biting but incisive commentary. After Ton-That was linked to ViddyHo's web domain registration, Valleywag posted a risqué photo of him sucking a lollipop under a headline that borrowed his tongue-in-cheek Twitter bio: "Was an 'Anarcho-Transexual Afro-Chicano' Behind the IM Worm?"

Valleywag called the "the classically quirky" Ton-That "a screaming stereotype of San Francisco's Web crowd" and mocked his track record of chasing the latest technological trend. "That's the irony of Ton-That's involvement with ViddyHo," the post said. "If he is indeed the perpetrator of the worm, it may make him hated. But it would be the first truly original thing he's done."

People recognized Ton-That on the streets of San Francisco in the days after the post ran. He was angry, feeling he'd been mischaracterized. Valleywag had reached out for comment, but Ton-That had ignored it, not understanding how the media work and hoping the whole thing would just blow over. "It was just a way to share videos," Ton-That would say later. "It just kind of went out of control."

A Valleywag follow-up post a month later said the police had been looking for Ton-That but that they had "never nabbed the hacker," because he was back with a "new website" that was simply ViddyHo under a new name. An FAQ attributed the previous site's problems to "a bug in our code."

Ton-That was twenty years old at the time. The incident destroyed his online reputation, and the Valleywag posts, specifically, would haunt his Google results for years to come. Burned by going it alone, and wanting to be part of a team, he took an engineering job at Ravikant's AngelList and then became an early employee at a fitness startup called FitMob. The benefits included equity and obscurity.

In 2015, around the same time Trump descended an escalator to announce his run for the presidency, FitMob was acquired by ClassPass, and Ton-That, as an early employee, got to cash out. He decided it was time for a change from San Francisco.

"Every party you go to there," he said, "it's like 'Hey, what do you do? What's your valuation? What are you working on?' I just got really tired of it."

Ton-That was at a crossroads. He had gotten divorced. He was frustrated professionally. He had spent almost a decade in the Bay Area, but his only lasting achievement was the eradication of his Australian accent. He wondered whether he should give up on the tech industry altogether.

He uploaded a few recordings of himself playing guitar to Spotify, including a hauntingly lovely melody he'd composed called "Tremolo." He chose as his digital album cover a shirtless photo of himself, splashed in a red substance that looked like blood, lighting a cigarette. He was trying on different identities, trying to find the one that fit best. He traveled the world for a few months, visiting Paris and Berlin, Vietnam and Australia. He attended Burning Man, an annual festival in

Black Rock Desert, Nevada, which attracts bohemians, creatives, and Silicon Valley types who consume art, music, and drugs as a form of soul-searching. Someone there took a serene-looking photo of him on a white bicycle, wearing a silky golden tunic and cream-colored pants, holding an oil-paper umbrella to shield himself from the bright desert sun; yet he had been a polarizing figure in his Burning Man camp. One campmate described him as charismatic and witty, while another said he was "a horrible human being" who loved to provoke and rile sensitive people. In other words, he liked trolling.

He eventually moved to New York City and told people, half jokingly, that he planned to become a model. But when someone he met at a party actually asked him to do a modeling gig, he agreed and discovered he hated it. So he stuck with tech. He released a handful of forgettable apps, many of which incorporated a new iPhone developer tool from Apple that made it possible to find a person's face in photos. There was TopShot, where strangers could rate your appearance on a scale of 1 to 5, and Flipshot, which would mine your camera roll for unshared photos and prompt you to send them to friends. "All the best photos of you are on other peoples' phones," he once mused in a tweet.

Two of the apps revealed a new political bent. When he released a Twitter clone called Everyone in April 2016, the promo art included a sultry photo of Ton-That in a red Make America Great Again hat. "Come to Tompkins square park, it's beautiful out and I have my guitar," said a message below it. And there was Trump Hair, which let a user add "The Donald's" signature coif to their photos. "It's gunna be YUGE!" promised the tagline. (It wasn't.)

Ton-That befriended technophiles like himself, but many of them leaned left. He wasn't entirely happy about that, tweeting in early 2016, "Why are all the big cities in the USA liberal?" On Twitter, he was suddenly all in on the alt-right, retweeting celebrities, journalists, and publications affiliated with libertarianism and white supremacy who expressed frustration with the current climate of political correctness. "I'll stop using stereotypes just as soon as they stop being right" was one sentiment that Ton-That deemed worth retweeting. He showed up at parties in a big white fur coat and his MAGA hat, and casually dropped extreme views into conversations: Chelsea Clinton's wedding, he said, had been an opportunity to launder money; race and intelligence are linked ("look at the Ashkenazi Jews"); "There are too many foreign-born people in the U.S.," he would say, according to friends who found those opinions confounding for many reasons, one of which being that he himself was an immigrant.

One of Ton-That's best friends from San Francisco stopped talking to him. "We couldn't find common ground. It had never happened to me before that I had ended a friendship that close," said Gustaf Alströmer, who was a product lead at Airbnb at the time. He said that Ton-That had not been a partisan person when he lived in the Bay Area. According to Alströmer, Ton-That's political interests had been more theoretical, about the fairest way to do voting or the best economic system for society. "He talked about racism in Australia and his father being an immigrant. He was thoughtful and had empathy and sympathy for people who came from the same background," he said. "The words he used in 2016 contradicted everything he said before." He attributed Ton-That's change to falling in with the wrong crowd online. "You find people who agree with you on Slack and Reddit," he said. "He was a person looking for an identity."

Like Alströmer, many of Ton-That's friends in New York couldn't understand how he had come to identify with the nationalistic, pro-white MAGA crowd. Ton-That was smart and sensitive, a feminist who played guitar soulfully at parties and had wandered the radically inclusive camps of Burning Man just the summer before. At the same time, he was a contrarian who liked to play the devil's advocate in an argument. His friends thought that was part of the reason he supported Trump. Donning a MAGA cap in Manhattan was a very easy way to get a rise out of people. "In today's world, the ability to handle a public shaming/witch hunt is going to be a very important skill," he tweeted presciently.

IN CLEVELAND, AT the 2016 Republican convention, Ton-That stuck out in a different way: He was an eccentrically dressed, long-haired hipster amid a drab khaki-and-button-down crowd. But he was there with someone who both looked the part and seemingly knew how to navigate this world: a paunchy, redheaded lightning rod named Charles Carlisle Johnson, better known on the internet as Chuck Johnson.

Johnson's name is rarely invoked without the word "troll" shortly thereafter, not because of his shock of rust red hair or his fulsome beard but because of his dedication to the online art of needling liberals. Johnson, twenty-seven, described himself at the time as an investigative journalist with high-profile conservative connections. He had written a book about President Calvin Coolidge that had gotten blurbs from conservative firebrand Ted Cruz, George W. Bush's torture defender John Yoo, and the media demagogue Tucker Carlson.

In 2014, after a summer fellowship at *The Wall Street Journal* and a few years reporting for Carlson's Daily Caller, Johnson launched his own site, GotNews. He had gained notoriety there for, among other things, his theories about Michael Brown, a Black eighteen-year-old shot and killed by a police officer in Ferguson, Missouri. Brown's death sparked a national uproar amid a building Black Lives Matter movement. The officer claimed that the unarmed Brown had posed a threat to his life. In defense of the officer, Johnson published photos from Brown's Instagram account that he said suggested a violent streak, and sued unsuccessfully to get Brown's juvenile court records. That was all chum in the alt-right water, and it worked. GotNews got views.

Johnson's investigations frequently targeted Black men. In 2015, Johnson was banned from Twitter after saying he wanted to raise money to "take out" Black Lives Matter activist DeRay McKesson. He claimed to have proof that Barack Obama was gay and that Cory Booker, who was then the Democratic mayor of Newark, did not actually live in Newark. The Southern Poverty Law Center called GotNews "an outlet for conspiracies." It was derided by the mainstream media and by Gawker, who dubbed Johnson the "web's worst journalist."

Occasionally, Johnson's hunches were correct, such as the time he questioned the veracity of a *Rolling Stone* article about a brutal gang rape at the University of Virginia, a story that fell apart under further scrutiny. But his methods were extreme: He offered cash to anyone who could out the anonymous alleged victim in the story. Johnson did eventually identify her, but then accidentally posted a photo of the wrong woman, who was then harassed. He would later say he regretted the lengths he had gone to at GotNews, blaming the "intemperance of youth" as well as "mimetic desire," a theory developed by René Girard, a French philosopher popular with the Silicon Valley set. In Johnson's interpretation, enemies cannot help but mimic one another. If he became vicious and cruel, it was only to reflect the Gawker bloggers he despised.

Johnson later claimed that he had done the muckraking at the behest of the government, saying he was an FBI informant tasked with drawing out domestic extremists by saying crazy things online. Riveting in a supervillain sort of way, shamelessly blunt with a sharp mind that made surprising connections, he was a relentless self-promoter and networker whose contacts list contained the numbers of billionaires, politicians, and power brokers. Whether they would actually pick up when he called was another matter; he didn't always keep friends for long.

When Charles Johnson and Hoan Ton-That decided to attend the Republican convention together, they had only recently met. Ton-That was a GotNews reader and asked to be added to a Slack group Johnson ran for extreme right-wing types. After bonding online, they decided to meet in person when Johnson came to New York.

"I loved Hoan from the second I met him," Johnson said. "He showed up dressed in this flowery suit thing, made in Vietnam. He wore the most outlandish, craziest shit. He's not like the other kids in the sandbox."

They had dinner with Mike Cernovich, a men's rights advocate who had used his considerable clout on Twitter to boost the #PizzaGate conspiracy theory alleging child sex trafficking by Democrats in the nonexistent basement of a D.C. pizzeria. Cernovich tweeted a photo of his dinner mates making the controversial "OK" hand gesture, an "all is well" signal co-opted by conservatives to communicate tribal solidarity and infuriate liberals. "Chuck Johnson and @hoantonthat say hi," Cernovich captioned it. He couldn't tag Johnson because he was still banned on Twitter.

Johnson and Ton-That's first meeting lasted ten hours. Johnson said they had discussed psychology, genetics, and how technology could expand what is knowable about human beings. Johnson felt that the Victorian Era was when global peak IQ was reached and that important lines of inquiry from that earlier age had been unfairly "canceled," such as physiognomy, the ability to discern the character of the mind from the features of the face. Though it was generally regarded by the academic community as racist and naive, Johnson thought scientists should reexamine physiognomy with the computing tools of the modern age. Because he believed that antiquity was the route to enlightenment, Johnson resolved at one point to read only books written before 1950. Ton-That later enthused to a friend that Johnson was brilliant.

Johnson had organized a group house in Cleveland for the Republican National Convention and was psyched to attend what he considered to be his generation's Woodstock. He said that Ton-That could crash with him. A few days before the event, Ton-That emailed Johnson asking if the offer still stood.

"I'll bring my guitar," Ton-That said.

"Want to meet Thiel?" Johnson offered, meaning Peter Thiel, the philosopher-lawyer turned venture capitalist, who had co-founded PayPal and backed Facebook in its early days, accumulating fabulous wealth along the way. Thiel was a libertarian who believed that freedom and democracy were fundamentally incompatible, that death could be overcome, and that the pursuit of technological progress and any resulting profits was a raison d'être. A powerful, antidemocratic billionaire who might live forever; what could go wrong?

"Of course!" Ton-That responded.*

A gender studies professor who owned the three-bedroom house in Cleveland where they stayed remembers Ton-That and Johnson. It was the only time she ever rented out her home, unable to resist the laughably high rates being paid by out-of-state Republicans. The group turned out to be respectful guests, but she had been worried enough about hosting Trump supporters that she removed an Andy Warhol

*Ton-That would later disavow the radically conservative views he expressed during this period, his time as a Trump supporter, and his friendship with Charles Johnson, but Johnson kept extensive documentation of their relationship in the form of emails, photos, and videos.

print of Chairman Mao Zedong from her living room wall. Her bookshelves, however, remained stocked with radical feminist literature, which amused her temporary tenants. In one photo from the week, Johnson sat in an armchair in his boxers, feigning enlightenment as he leafed through one of the tomes.

Ton-That followed through on his promise to bring his guitar and frequently serenaded his friends. In one smartphone video Johnson captured, Ton-That performs an original song he composed, inspired by Milo Yiannopoulos, a right-wing British provocateur who'd been kicked off Twitter that week for harassing a Black actress.

"You're everybody's favorite gay dude," Ton-That sang. "And you're always throwing grenades / Over to the other side / And they can't survive / So what they do is / They ban you."

Their Airbnb was a fifteen-minute drive from the Cleveland Cavaliers' home base, which had been temporarily transformed from a sports arena into a political coliseum. Huge photos of NBA superstar LeBron James, a vocal critic of Trump, had been covered with Republican elephants. On the first night of the convention, Ton-That and Johnson took an Uber to the arena to hear Melania Trump speak. They sat in the nosebleed seats and took her in mainly on the huge screen above the stage.

The next morning, Johnson followed through on the introduction he had dangled, taking Ton-That in an Uber to an "epic house" in Shaker Heights to meet Peter Thiel, who had just rolled out of bed. His private chef fixed them a superfoods breakfast of eggs with salmon and avocado that they ate poolside. The main topic of conversation, according to Johnson, was Gawker. Thiel had hated the media company for many years by that point. It had covered him frequently, from headlining his sexuality—"Peter Thiel is totally gay, people"—to criticizing his "loopy libertarian" views, including a dream of building a floating city outside of any government's jurisdiction and the belief that letting women vote had been bad for American democracy. The outlet had also closely covered the travails of his investment fund, Clarium Capital, which had lost billions of dollars before shutting down.

"I think they should be described as terrorists, not as writers or reporters," Thiel once said of the site's journalists. The billionaire had been secretly funding lawsuits against the company, hoping to bankrupt it, and just the month before he met with Ton-That and Johnson, one of them had succeeded.

A former professional wrestler, Hulk Hogan, had sued Gawker in Tampa, Florida, for publishing nearly two minutes of surveillance-cam footage that featured him having sex with a friend's wife in a 2012 post headlined "Even for a Minute, Watching Hulk Hogan Have Sex in a Canopy Bed Is Not Safe for Work but Watch It Anyway." Gawker assumed it would be protected by the First Amendment, America's strong constitutional shield for speech, because, as it argued in court, the content of the post was accurate and of public interest. Gawker's employees hadn't stolen the tape, which would have been a crime; they were only reporting on and sharing its existence. American courts were generally protective of the media, but even if it were a challenging legal battle, Hogan had little to lose; his dignity had already taken the hit and he was backed by a billionaire. As it turned out, Thiel's investment paid off: Hogan body-slammed Gawker. A jury in Florida awarded him $140 million, leading the media company to file for bankruptcy. A victorious Thiel then admitted his secret role in the case, sending chills down the backs of journalists across the country. The powerful can't be held accountable if they can simply sue critics out of existence.

According to Johnson, Ton-That became emotional talking with Thiel about how much the blow against Gawker meant to him, given the way the company had damaged his own reputation. Johnson, too, had joyful tears in his eyes. He had sued Gawker a few years earlier for defamation, and while the company was winding down, it would settle his lawsuit for, according to Johnson, more than a million dollars.

Two days later, when Thiel gave a speech at the Convention, Johnson and Ton-That were no longer in the rafters but on the main floor of the arena, close to the stage, having obtained VIP passes. They took a giddy photo, smiling, with Thiel's face projected onto the massive screen behind them. Thiel's visage reflected his German heritage—a broad forehead, heavy brows—and a youthfulness at forty-nine that seemed to lend credence to rumors he was being injected with the blood of young people.

Seeming unaccustomed to such a large crowd, Thiel stared fixedly at three spots in the arena, swiveling his head robotically at the end of each clause—left, center, right . . . left, center, right—to make eye contact with an audience who likely didn't know who, or how powerful, he was. He talked about how America was falling behind, how it was broken, and how a businessman like Trump was the one to fix it. Trump gave his own ominous speech later that night, invoking an America beset with violence, crime, and illegal immigrants and warning that his opponent, Hillary Clinton, would bring more of all three.

Johnson said that his and Ton-That's fandom of Trump was "absurdist," and that they appreciated how he livened up the typically boring political world with his antics. "If you're a weird person and you see a weird man run for president, you're like 'Ah, I finally have a seat at the table. The weirdos are in control,'" Johnson said. "We got the joke of it."

Who they truly respected was Peter Thiel and his call for tools to help "fix" America. Johnson and Ton-That chatted that week about how useful it would be to have a tool to pull up information on the other attendees at the convention, to be able to determine whether a stranger there was worth knowing or a foe to be avoided.

One night, after they got out of an Uber in front of their Cleveland Airbnb, Johnson asked Ton-That if he wanted to do more with his life than make small-time apps. "Do you want to do something big?" Johnson recalled asking him. Johnson remembers that Ton-That looked back at him with a wolfish grin and a trickster gleam in his eye and said, "This is going to be a lot of fun. Of course I'm in."

And that was when their journey to unlock the secrets of the human face began. It was an ambition with very deep roots.

CHAPTER 2

THE ROOTS (350 B.C.–1880s)

More than two thousand years ago, Aristotle declared that man was the only creature on the planet with a true face. Other animals had the basic components—eyes, a mouth, and a nose—but those organs alone did not suffice. A true face, according to the Greek philosopher, conveyed personality and character. It was a reflection of the human soul with all of its depth and nuance. Yet it was apparently simple enough to decode that Aristotle could assert with confidence that men with large foreheads were lazy while those with fatty tear ducts were liars.

"When men have large foreheads, they are slow to move; when they have small ones, they are fickle," he wrote in his *History of Animals*. "When they have broad ones, they are apt to be distraught; when they have foreheads rounded or bulging out, they are quick-tempered."

His list was lengthy. People with straight eyebrows were sweet, but if the brows were arched, the person would be cranky, funny, or jealous, depending on which way they curved. People with green eyes were cheery, and people who blinked a lot were indecisive. People with big ears would talk your own off. And on and on. (He also said that flat-footed men were "prone to roguery.")

It was palm reading but for the face. The so-called science of judging a person's inner self from their outside appearance came to be called physiognomy. Though Aristotle's judgments seemed to be pulled out of thin air, they intrigued serious thinkers in the centuries that followed, particularly in the Victorian Era, when an English polymath named Francis Galton took the measure of man to even darker places.

Galton, born in 1822, the youngest of seven children, belonged to an esteemed family. His grandfather was Erasmus Darwin, a renowned doctor, inventor, poet, and published author. His older cousin was Charles Darwin, whose theory of evolution would shake the scientific world. As young men, Darwin and Galton were both pushed to become doctors, but Darwin instead took up the study of plants, beetles, and butterflies. When Darwin graduated from Cambridge in 1831, his professor recommended he join the crew of the HMS *Beagle* as the ship's naturalist. Darwin was very nearly rejected from the five-year voyage that yielded his groundbreaking theories because the ship's captain was an amateur physiognomist who didn't like the cut of Darwin's jib.

"He doubted whether anyone with my nose could possess sufficient energy and determination for the voyage," wrote Darwin in his autobiography. "But I think he was afterwards well-satisfied that my nose had spoken falsely."

When Darwin returned from the other side of the world to write his first book, *The Voyage of the Beagle,* Galton was half-heartedly pursuing medical school, put off by the grisly autopsies and macabre operations performed without anesthesia. After his father died in 1844, Galton dropped out, but, unsure what he should do next, he consulted a phrenologist.

Phrenology, a "scientific" cousin of physiognomy, was based on the belief that a person's character and mental abilities could be read from the skull's shape and its unique lumps and bumps. The phrenologist's job was made a little easier because Galton, despite being just twenty-seven, was already balding like his cousin, with dark, thick sideburns. After thoroughly massaging Galton's head, the phrenologist wrote a seven-page report in which he explained why the failed medical student was best suited for "roughing it," as a colonizer or by serving in the military.

Galton resolved to become an explorer and spent two years traveling in an area now known as Namibia, in Africa, by horse, mule, and wagon, with nearly forty companions, shooting giraffes, rhinos, lions, and zebras for food and fun. Upon his return, he, like his cousin, wrote a book about his travels: *Narrative of an Explorer in Tropical South Africa.*

Galton's tales of hunting and exploring were, in keeping with the time and place, viciously racist. He referred to the locals as "savages," describing one ethnic group as "warlike, pastoral Blacks"; another as "intelligent and kindly negroes"; and a third as having "that peculiar set of features which is so characteristic of bad characters in England . . . the 'felon face.'" "I mean that they have prominent cheek bones, bullet shaped head, cowering but restless eyes, and heavy sensual lips," he wrote. Galton's adventures into "the unknown" brought him name recognition and awards from the Royal Geographical Society.

"What labours and dangers you have gone through," his cousin Charles Darwin wrote to him in 1853, shortly after the book was published. "I can hardly fancy how you can have survived them, for you did not formerly look very strong."

It was a congenial ribbing between cousins and a reflection of a relationship that, on Galton's side at least, inspired competition. In the wake of Darwin's publication of *On the Origin of Species* in 1859, Galton searched for his own way to make contributions to the scientific world. His epiphany was to apply Darwin's ideas about how populations evolve to human beings, and to search for evidence that spiritual and mental traits could be inherited along with physical ones. To assess whether "talent and character" were passed down in families, he looked at the pedigrees of "eminent" individuals, listing their relations and their accomplishments. He found, based on his own subjective definition of "eminence," that the descendants of distinguished men were more likely to make significant contributions to society. He and Charles Darwin, "the Aristotle of our days," he wrote, were proof of the theory, descending as they did from the inimitable Erasmus. Galton gave little consideration to the role of societal status, class, and the greater spectrum of opportunity that exists for the offspring of rich and famous people—nor to the fact that Darwin's genius had almost been thwarted by his bulbous nose.

Seeking to push his ideas further, Galton superimposed photographs of criminals—sorted into rapists, murderers, and fraudsters—on top of one another, hoping to surface physical traits shared by each category of criminal. In his view, people born with certain traits were destined to follow a particular path: Intellectuals and criminals alike were born rather than made.

Darwin had found unique traits that emerged in isolated populations of animals on the Galápagos Islands, and Galton argued that the same was true of populations of races. He wrote, based primarily on other people's "observations" of them, that people of West Africa had "impulsive passions, and neither patience, reticence, nor dignity." Native Americans were "naturally cold, melancholic" with a "sullen reserve." The new white Americans were "enterprising, defiant, and touchy" and "tolerant of fraud and violence," because their forebears who had left Europe to settle in the new continent were restless sorts. He also ranked the races' intelligence. These were all inherited traits, he confidently claimed, passed from parents to their children.

Galton thought that society should seek to perpetuate the best traits of humankind, like farmers trying to breed the hardiest plants and healthiest animals. He encouraged the marriage of the most able men and women to improve the quality of the human race overall. He coined the term *eugenics* from the Greek word *eugenes,* meaning "good in birth." As for the less able, Galton later wrote that "habitual criminals" should be segregated, surveilled, and prohibited from having children. "It would abolish a source of suffering and misery to a future generation," wrote Galton, whose theories were embraced by academics in those future generations and by the Nazi regime in Germany.

Charles Darwin wrote to Galton after his book *Hereditary Genius* came out in 1870, admitting to having read only about fifty pages of it. He called it "interesting and original" but challenging to digest. "I have always maintained that, excepting fools, men did not differ much in intellect, only in zeal and hard work," he wrote, suggesting a deep disagreement with the premise of the book in an otherwise extremely polite letter.

Reviews of the book were mixed, with some reviewers calling Galton an "original thinker" and others mocking his overconfidence in his arbitrary theories. Galton's photo composites and mining of family trees were ostensibly attempts to measure the human mind and soul. But instead of providing scientific rigor to the study of human nature, he wrapped bias and prejudice in the trappings of high-minded analysis.

GALTON WAS JUST one of many Victorians trying to bring the scientific method to the study of the human body. In France, a surly records clerk named Alphonse Bertillon was taking a different approach, seeking to discover what made individuals distinct, rather than the same, using methods that would come to be widely adopted by law enforcement agencies.

Bertillon was tall and haughty, with large ears, a bushy beard, and fierce, dark eyes. Born in Paris to a family of statisticians, he got a job, at twenty-six, at the police department. It was dismal. Bertillon worked in the basement copying information from arrest reports and background checks onto criminal identification forms. France had stopped branding convicted criminals at the beginning of the century and was keeping track of them instead via written descriptions on alphabetically arranged cards and with black-and-white cameras, a relatively new invention. But it was a terrible system. The accused would simply make up fake names, so the police couldn't reliably check the card catalog to see whether someone was a repeat offender.

Bertillon was an order obsessive who yearned to better organize these convicts. So he visited prisons and used calipers to take inmates' measurements: their heights; the lengths of their limbs, hands, and feet; the breadth of their heads; the distance between their eyes; the distance from the cheekbone to the ear, and so on.

Bertillon determined that each person was unique, his particular combination of measurements a reliable way to sort one from another. He proposed a system to his boss at the police department: Take eleven measurements of each criminal's body and use that information to keep track of society's bad elements, sorting arrested persons into categories of physical specimens, starting with the length of the head—small, medium, or large—then its breadth, then the distance from cheek to ear, and going on from there.

His boss thought he was insane. A few months later, Bertillon got a new boss, who was still skeptical, but approved a trial to see whether the system could truly identify a repeat offender. Within three months, thanks to Bertillon's system, a person brought into the station for theft was identified as a recently released prisoner. More successes followed. The system was instituted in Paris in the 1880s and used for years, eventually spreading to law enforcement agencies in other countries.

Bertillon became a renowned criminologist who also pushed to standardize the photographing of people hauled in for misbehavior: two photos, one from the front and a second in profile; the modern mug shot.

In 1893, Francis Galton, intrigued by a system for tracking criminality, traveled to Paris to visit Bertillon's laboratory and got a souvenir, a personalized card with his measurements and mug shots, his long sideburns gone white. But Galton, like police officers themselves, saw drawbacks to Bertillon's meticulous measurements for the criminal justice system: It was challenging to measure an uncooperative subject, and different people took measurements in different ways.

Galton had heard about a British colonial officer in India who had become an evangelist for fingerprinting, and collected prints from whoever was willing to give them. When Galton got wind of the collection, he asked to study it. Galton hoped to find patterns of intelligence and evil in the whorls and ridges, but he was sorely disappointed. He could find no fundamental differences in fingerprints across ethnicities or between an eminent statesman and a "congenital idiot." "No indications of temperament, character or ability can be found in finger marks, so far as I have been able to discover," he wrote.

Bertillon thought his own methods were superior, but after Galton wrote a book about how fingerprints could be useful for criminal investigations—a piece of evidence unwittingly left behind at the scene of a crime—Bertillon reluctantly added room for them on his convict cards. After two prisoners in the United States were mistaken for each other in 1903 because they had similar body measurements and the same name, fingerprints became law enforcement's preferred approach.

Bertillon did not agree with Galton that nefarious types were anatomically different from upstanding citizens, but Galton did have other intellectual company among his contemporaries. In 1876, an Italian physician and phrenologist named Cesare Lombroso published a book called *Criminal Man* dedicated to identifying deviants. It described criminals as if they were their own species, with distinct physical and mental characteristics.* A large jaw, a thin upper lip, protruding mouth, or the inability to blush were all damning features, as were bodily characteristics Lombroso described as "found in negroes and animals" and certain cranial bones "as in the Peruvian Indians."

Some serious thinkers at the time were worried about atavism, the idea that humankind was devolving. Lombroso suggested that criminals demonstrated this degeneration and that they had the "physical, psychic, and functional qualities of remote ancestors." In other words, if someone on the street looked like a "savage," women should tightly clutch their purses. It was, again, racism masquerading as scientific rigor, a line of thought that seemed dangerously entwined with the desire to measure and classify the human form.

As outrageous as this sounds, Lombroso was taken seriously as a criminal anthropologist by law enforcement officials, who called him in to consult on unsolved cases. After a child was sexually assaulted in the late 1800s and infected with syphilis, police asked Lombroso to review the suspects. He identified one as the culprit based on his "sinister physiognomy," Lombroso wrote later, as well as traces of a recent attack of syphilis.

His theories were influential. Charles Ellwood, who would become the president of the American Sociological Society, gushed after Lombroso's book was translated into English that it "should mark an epoch in the development of criminological science in America." In 1912, Ellwood wrote in the *Journal of the American Institute of Criminal Law and Criminology* that "Lombroso has demonstrated beyond a doubt that crime has biological roots."

This belief that all things could be measured, even criminality, was a reflection of an embrace of statistics, standardization, and data science in the nineteenth century. This fervor was shared by industry—particularly American railroads, which became enthralled by the

* Lombroso's daughter wrote in the introduction to a reprint of his book that he had once presented forty children with portraits of great men and of criminals, and 80 percent of the children were able to pick out the "bad and deceitful people," proof apparently of his theory and not of a difference in portraiture styles.

possibility of sorting and tracking their customers. Railways employed a rudimentary facial recognition system for conductors: tickets that featured seven cartoon faces—an old woman, a young woman, and five men with varying styles of facial hair. The conductor would mark each ticket with the assigned face closest to the ticket's holder, so that if an elderly woman bought a round-trip ticket, a young man would not be able to use it to make the return journey.

Another more complicated system for sorting passengers left many of them disgruntled, according to the August 27, 1887, edition of *The Railway News:*

> Much complaint has been made in consequence of the introduction of a new ticket system on overland roads to California. The trouble all arises from the fact that the ticket given at the Missouri River contains what is called a "punch photograph" of the holder. This is supposed to be a complete description of the passenger. Along the margin of the ticket is printed, in a straight column, the following words in small, black type:
>
> Male—Female.
> Slim—Medium—Stout.
> Young—Middle-aged—Elderly.
> Eye. Light—Dark.
> Hair. Light—Dark.
> Beard. Moustache—Chin—Side—None.

A train agent "photographed" a passenger by punching out all the words that didn't describe him or her. The article noted that a woman labeled "stout" and "elderly" might take issue with the system but reported, more seriously, that it had caused problems for travelers trying to complete a round trip who were deemed not similar enough to their ticket description.

This analog facial recognition system inspired one important traveler: a son of German immigrants named Herman Hollerith. Hollerith, a former U.S. Census Bureau worker, thought the "punch" method could assist in that ambitious collection of demographic data from Americans across the country. He came up with the idea of cards with standardized perforations that could be read by an electromagnetic machine that looked like a waffle maker filled with pins. When the machine was closed on a card, the holes would allow pins through so they could make contact with a well of mercury, completing an electrical current. That circuit energized a magnetic dial tabulating the data, advancing a counting hand. If the pin dropped through the hole designated for a married person, for example, the circuit was closed and that person was counted as married. No hole: no circuit connection and no matrimonial bliss.

When put to the test, Hollerith's system shaved many hours off bureau clerks' tabulation of data, and after it was successfully used for the 1890 census, other census bureaus around the world leased his machine and bought his punch cards. It was one of the first modern computers, and Hollerith went on to found the company now known as IBM.

In the early twentieth century, an anthropologist at Harvard University named Earnest A. Hooton used IBM punch cards and the Hollerith card sorter to pick up where Galton, Bertillon, and Lombroso had left off. Starting in the late 1920s, Hooton took the measurements of 13,873 criminals and 3,203 civilians across America. In *Crime and the Man,* a more-than-four-hundred-page book about the study, published in 1939, he concluded that Lombroso was right. "Crime is not an exclusively sociological phenomenon, but is also biological," he wrote. He waffled on whether it was possible to finger a potential offender on looks alone but, like Galton, suggested that repeat offenders should be "permanently incarcerated and, on no account, should be allowed to breed." He believed that with more study, it would be possible to determine "what types of human beings are worthless and irreclaimable, and what types are superior and capable of biological and educational improvement."

"We can direct and control the progress of human evolution by breeding better types and by the ruthless elimination of inferior types," the Harvard anthropologist wrote. "Crime can be eradicated, war can be forgotten."

The field of eugenics, with its idea that certain types of people were poisonous to society, was infamously adopted in Nazi Germany. That it was prevalent in America as well gets less attention in history textbooks. Native American and Black women and the "feeble-minded" were sterilized against their will in the United States as recently as the 1970s. Sterilization was justified, according to a chilling 1927 Supreme Court ruling, because "society can prevent those who are manifestly unfit from continuing their kind."

Crime and the Man is not in wide circulation today. Even at the time, critics called Hooton's theories "highly unorthodox," "controversial," and "startling." But this strain of thought—that criminality and genius are biological, that who someone is can be read on their face—persisted, as did the belief that computers, with their vast abilities to process data, are capable of uncovering who we really are.

CHAPTER 3

"FATFACE IS REAL"

After the Republican National Convention in Cleveland, Hoan Ton-That returned to New York City and Charles Johnson to Clovis, California, outside of Fresno. Being three thousand miles apart wasn't a problem; they had the internet and, according to Johnson, a vague idea about what they might create together: a tool to help make decisions about strangers.

Because two guys in their twenties might not be taken seriously by investors, Johnson thought they needed a third person with some gravitas—much like in the early days of Google, when Eric Schmidt had been brought in to provide "adult supervision" for the company's young founders. Johnson even had a candidate for the job: Richard Schwartz, a longtime New York City politico who had once worked for Mayor Rudy Giuliani.

Schwartz grew up on Long Island, New York, moved to New York City to attend Columbia University, and then never left. After he graduated in 1980, he worked for the Department of Parks and Recreation, a formidable city agency with a huge budget. It was run at the time by disciples of Robert Moses, a polarizing public official credited with beautifying parts of New York City with parks and playgrounds who also, notoriously, ran highways through the middle of minority neighborhoods, leaving them ugly and polluted.

"I was taught deliberately many of the trade secrets of how to get projects done in government," Schwartz said. "Because that's what Moses was about."

In 1994, when former prosecutor Rudy Giuliani became the first Republican mayor of New York City in decades after running a "law and order" campaign, he appointed Schwartz to be a policy advisor. Schwartz worked on the privatization of city services and on welfare "reform," trying to reduce the number of people getting government benefits by instituting work requirements. At the end of Giuliani's first term, Schwartz went private himself and launched a lucrative consulting business, called Opportunity America, that helped other cities cut down on their welfare rolls, leading one local journalist to point out that a "system that once gave a lot of money to poor people has been replaced by one that gives a lot of money to the people who put the poor to work." Soon after, his firm got mired in a New York City corruption scandal, with allegations that Schwartz had used his mayoral connections to help a Virginia-based company land a $104 million job-training contract—a chunk of which was earmarked for Opportunity America. The controversy eventually blew over, but Schwartz sold his firm to that Virginia company for nearly a million dollars and got out of the welfare consulting business.

Then he made a surprising career move, landing a job in early 2001 as the editorial page editor at the New York *Daily News.* This led to more critical stories about Schwartz by media reporters who were perplexed by his lack of obvious qualifications. "Mort liked me," he said of Mort Zuckerman, the billionaire real estate magnate who owned the paper, to whom he was introduced by Giuliani contacts. Schwartz oversaw editorial coverage of the terrorist attacks on the World Trade Center, including an ironic push, given his past work in privatization, for the federal government to take over airport security, presaging the creation of the Transportation Security Agency (TSA).

By 2006, he had left the *Daily News* to go into the behind-the-scenes world of management and strategic consulting, giving his advice and connections to whoever was willing to pay for them. His partner was a longtime PR guy named Ken Frydman; they had worked together for Giuliani and then for the *Daily News.* They parted ways after five years, Frydman said, because Schwartz had allegedly tried to cut him out of a financial deal. "He was my partner, but in the end, I couldn't trust him," Frydman said. "I do miss him. We made good money. We had good times."

Schwartz got married and had a daughter, but by 2015, when he had lunch with the redheaded conservative upstart Charles Johnson, he didn't have much else going on.

SCHWARTZ AND JOHNSON were both supreme networkers, so it's not surprising that they would cross paths, but it was an unusual pairing. Schwartz, a baby boomer in a boxy, ill-fitting suit, had come to appreciate the protection that came from keeping a low profile, while the brash millennial Johnson had achieved his power by trolling the internet.

They met under a bridge at Schwartz's go-to lunch spot: Pershing Square, an old-fashioned diner across from Manhattan's Grand Central Station. The entrance was shadowed by Park Avenue, which ran overhead; Schwartz said he had gotten the unusual location for a restaurant approved when he worked for the city. It was ten months before Ton-That and Johnson met, and a month after Trump had announced that he was running for president.

"Richard wanted some reintroductions to Trump world. I had some. We compared notes on billionaires and real estate people in New York," Johnson said, describing the meeting. "I collect superrich people. It's a thing. Richard had some and I had some, and we were basically swapping them like baseball cards."

Schwartz told Johnson his tales of life inside the Big Apple's political machine and about his time working for Rudy Giuliani. For Johnson, it was like visiting a political museum, dusty but informative. As he left the lunch, he filed Schwartz away in his brain as someone with connections to wealthy families and the Giuliani political tribe.

A year later, while Johnson was hatching a plan to do "something big" with Ton-That, who was based in Manhattan, Schwartz sprang to mind. In July 2016, two days after Peter Thiel's speech in Cleveland, Johnson was traveling home to his Indonesian American wife* in California, when he dashed off an email to Ton-That and Schwartz: "You should meet in NYC ASAP!"

They soon did, and though thirty years separated them, they hit it

* Johnson frequently mentions the ethnicity of the woman he married as proof that he is not a white supremacist.

off. Schwartz was fatherly and encouraging, qualities that resonated with Ton-That, whose own father, a professional gambler turned university professor, was half a world away. Schwartz said that Ton-That struck him as a "lost soul."

Schwartz talked about how he had transformed the East Village's Tompkins Square Park, where Ton-That liked to go play guitar. Schwartz had lived a block away from the ten-acre park in the 1980s when he was Ton-That's age and working for the Department of Parks and Recreation. Back then, it had been a hangout for the homeless, drug dealers, partiers, and leather-clad punk rockers who blasted boom-box radios late into the night. Schwartz wanted it to be a more orderly place with dog runs, where families could gather and young professionals could have a coffee. In 1988, the city started enforcing a curfew on the previously twenty-four-hour park, and when people protested, hundreds of NYPD officers were called in, resulting in an all-night brawl that left dozens injured. The police presence increased, with a push to remove the park's homeless camps, and in 1991 it closed for a year for renovations. That, combined with gentrification in the area, transformed it into the place Schwartz had envisioned, where residents sunbathed, families brought their kids to play, and a young lost soul in a MAGA hat could serenade his friends.

Ton-That asked him why he didn't move to San Francisco—the liberal-leaning tech mecca had rampant homelessness—and fix that place up. Schwartz laughed. "No way," he said. "I am never going there."

Ton-That told Schwartz he loved building, too, but in the virtual world, spinning his ideas into computer programs. Schwartz listened, intrigued, but thought that Ton-That needed help to dream bigger. They started brainstorming. "If he can code half as well as it sounds like he could do, we could do anything," Schwartz recalled thinking. "He was a classic character in search of an author."

Schwartz left the meeting unsure where it would lead. He liked the genius coder immensely but thought the younger man was "too cool" for him. So he was surprised when Ton-That called him just a few days later. "Hey, buddy!" Ton-That said.

BY SEPTEMBER 2016, Ton-That, Schwartz, and Johnson were emailing one another frequently, discussing a tool they could build that would make it easier to judge people. Their original idea was to enter someone's email address or social media handle, or even a photo of their face, into a search tool, which would return a report that accumulated everything available about the person online.

For Johnson, who prided himself on reexamining discredited sciences, such as physiognomy, for ideas worth salvaging, it was, unsurprisingly, the face that offered the most interesting possibilities. And he didn't have to rely on Victorian Age science. Modern-day researchers were seeking to prove the merits of physiognomy. Psychologists at Tufts University asked volunteers to guess people's sexuality based on cropped photos of their faces. They were right often enough that the academics suggested that "gaydar" was a real phenomenon. Another set of researchers did the same experiment with AI and said that it too could predict sexuality at rates better than chance. The suggestion that sexuality is written into our very facial structure was effectively dismantled by other academics who pointed out flaws in the studies, including a possible correlation between sexuality and personal grooming choices.

But there was more: Two Chinese computer scientists claimed to be able to predict a "criminal face" with 90 percent accuracy. Then, dueling American teams, one of which name-checked Cesare Lombroso as an inspiration, came up with their own computer programs for detecting deviants, claiming up to 97 percent accuracy. These so-called criminal detectors were also thoroughly savaged by critics, who said the programs weren't predicting who was a criminal, just which faces had been cropped from mug shots.

Schwartz seemed the wariest of going in that direction. He pushed his partners toward applications they could more easily monetize, such as security. When he sent them an article from *The Wall Street Journal* about companies using selfies as a password alternative, however, Ton-That was dismissive. "This is really dumb," he responded. "I can go to someone's Facebook and print out a photo of their face."

Ton-That aspired to do something greater. He was open to the idea that the face was a window into the soul, and that a computer might be trained to decipher it. "I do believe in genetic determinism," he said in one email, "so if criminality or other traits are heritable it should show up in the face." In messages he sent over the next few months, he suggested data crunching the faces of known murderers, cheaters, and drug abusers to find shared facial features that could potentially be used to predict whether someone might do you wrong. Johnson's beliefs had clearly influenced him.

Ton-That was the workhorse of the group. He combed the internet for materials to flesh out their ideas, searching for digital collections of faces and academic research papers that might offer other promising ways to analyze people. It was the entrepreneurial equivalent of throwing spaghetti at the wall to see what sticks.

"Fatface is real," Hoan wrote in one email to his collaborators about a 2013 study out of West Virginia University. Researchers had trained computers to predict a person's body mass index and by extension aspects of their health, based only on a photo of their face.

Late that same night, he sent another email with the subject "Face & IQ," including a link to a dubious 2014 study out of the Czech Republic. The researchers had gotten eighty student volunteers—forty men and forty women—to take an IQ test and have their portraits taken. Then another group of students looked at those portraits and hazarded a guess as to how intelligent each person was. Ton-That summed up the findings: "People could predict a man's IQ from a face but not a woman's." "Makes sense from evolution," he cryptically remarked, suggesting that only a man's intelligence matters for mating purposes.*

An automated tool able to judge people at a glance based on computer vision could be lucrative, the team seemed to think. Ton-That came across an article about a computer vision company called Clarifai that allowed people to take a photo of a product they liked, such as a pair of sneakers, and be shown similar versions for sale. The company had

*Years later, Ton-That said he deeply regretted sending emails described in this chapter and that they do not represent his "views and values."

just raised $30 million from investors. "People putting crazy money into this shit," he wrote in an email sharing the article with his partners.

"Impressive and encouraging," Schwartz replied.

But data mining people's inner selves required access to the kinds of secrets that are generally exposed only against a person's will, by reading a diary or by hacking into a private account. Security being an imperfect art, such breaches are not uncommon. In a revealing 2015 hack, so-called anticheating vigilantes exposed more than 30 million users of an extramarital dating site called Ashley Madison. The hackers got their hands on the site's billing records and posted them online, revealing the members' names and contact details. Ton-That described the breach as a gold mine of data on infidelity. "We could use email from Ashley Madison and look up their Facebook profiles and see what a cheating face looks like," he wrote in an email, evoking the spirit of Galton's "felon face."

"Love this," Johnson replied.

When a couple of Danish researchers scraped the dating site OkCupid and published the usernames of seventy thousand online daters, along with their answers to intimate quizzes, Ton-That saw more fodder for specious biometric data mining and suggested matching their faces to their "drug use and sexual preferences." He reached out to one of the technologists who had done the OkCupid scraping—a Danish blogger with ties to the far right, who unabashedly believed in a link between race and intelligence—and said he was "good to bounce ideas off."

"He suggested scraping inmates on death row," Ton-That wrote, including a link to a collection of mug shots of condemned offenders in Texas.

"Good stuff man," Johnson replied.

Some of Ton-That's friends worried about how much time he was spending with Charles Johnson, whom they considered to be a dark influence on him. When Ton-That visited California, Johnson took him to a shooting range. "If you've never fired a gun, I'm suspicious of you," Johnson said. He made a video of Ton-That fumbling with the bullets as he tried to load the weapon. "Guns rock," Ton-That said after the target practice.

The emails among Johnson, Schwartz, and Ton-That continued for months, spiraling closer to the tool they would eventually create. Ton-That, who had built a Tinder bot to automatically swipe right on everyone suggested to him to help maximize his dating opportunities, proposed building another Tinder bot to scrape the faces of everyone on the hookup app. Then, they could map them to their stated interests and likes, as if there were some telltale facial feature that revealed whether someone liked dogs or long walks on the beach.

But the team was not only thinking about predictions based on faces; they were starting to think about identifying them. One day, "if our face matching gets accurate," Ton-That wrote, you could use someone's photo to find their otherwise anonymous Tinder profile. It was a potentially dangerous idea: A stalker could find the object of his obsession on the dating app and make it seem like an algorithmic happenstance.

In late October 2016, Ton-That sent Johnson and Schwartz an article from the British newspaper *The Independent* titled "China Wants to Give All of Its Citizens a Score—and Their Rating Could Affect Every Area of Their Lives." A subhead explained, "The Communist Party wants to encourage good behaviour by marking all its people using online data. Those who fall short will be denied basic freedoms like loans or travel."

Most people sharing the article on social media seemed terrified and enumerated the risks posed by this kind of policy: It could mean that all of your mistakes would follow you everywhere you went; there could be no fresh starts. Your opportunities would be limited by judgments rendered about you in the past, correct or incorrect as they might be. One person said that it was something straight out of Gary Shteyngart's dystopian novel *Super Sad True Love Story.* Schwartz responded, "Amazing! Can we beat their 2020 deadline?"

A month later, Ton-That sent Schwartz and Johnson another ominous article about a new Russian face recognition app called FindFace that let people take photos of strangers and then find them on VKontakte, a social network like Facebook. A photographer had used the app to identify strangers on the subway. More nefariously, someone had used the app to find the real identities of porn actresses and sex workers and harass them. "This will be the future," Ton-That wrote.

It was a future that industry, the government, and computer scientists had been pushing toward for decades, a dream so enticing that few seemed to consider that it could become a nightmare.

CHAPTER 4

IF AT FIRST YOU DON'T SUCCEED (1956–1991)

Most Americans check their smartphones hundreds of times a day and realize within minutes if they have left them behind. Periodic surveys in the last decade have asked people what they'd rather sacrifice for a year: their smartphone or sex? In every survey, nearly half of the respondents chose abstinence over the loss of their touch screens. How did computers come to be such a dominant part of our everyday lives, so essential an appendage that some regard them as more necessary than a romantic partner?

The people who put us on the path to lives suffused with technology hoped, with an almost divine fervor, to create machines in our image that could not just augment our own abilities but provide true companionship. Many histories of modern computing point with reverence to the summer of 1956, when a group of brilliant mathematicians, engineers, psychologists, and future Nobel Prize winners gathered at Dartmouth College in New Hampshire. In a time of typewriters, black-and-white televisions, and fancy calculators the size of refrigerators, the all-male roster of attendees spent part of their summer vacation imagining machines that could think and act like humans.

"An attempt will be made to find how to make machines use language, form abstractions and concepts, solve [the] kinds of problems now reserved for humans, and improve themselves," said a proposal that convinced the Rockefeller Foundation to put up half the money requested. They wanted to make something like Rosie from *The Jetsons*—though the iconic cartoon wouldn't air until the next decade—a robot that could clean the house, do errands, take care of the kids, and crack jokes. They called it "artificial intelligence."

As heady discussions unfolded about thought and consciousness, two different ideas emerged about how to build machines that could imitate and even surpass humankind. The first idea, preferred by the logical, math-driven members of the group, was to give machines explicit instructions to follow, basically creating very long "If this, then that" rulebooks. Others thought that biology and psychology offered a more helpful guide and that a machine like this would need a kind of nervous system, something like a brain, with the ability to learn on its own through perception and experience. The first camp wanted to endow computers with human logic; the second wanted to give them human intuition.

Shortly after the conference, one of the organizers, a mathematician named Marvin Minksy, accepted a professorship at MIT. Just thirty-one but looking older with glasses and a bald pate, he quickly became a towering figure in the field. He experimented with both approaches to artificial intelligence, the logic-based version, called "symbolic AI," and the think-for-itself version, which was eventually dubbed "neural networks technology." He thought the former was more promising and co-wrote a book called *Perceptrons* that savaged the neural net technique, almost single-handedly cutting off government funding for that approach for decades.

In 1960, a few years after the Dartmouth gathering, Woody Bledsoe, a mathematician who had grown up on an Oklahoma farm, launched a technology research company on the other side of the country, in Palo Alto, California. The area was not yet known as Silicon Valley, but it was already at the center of the computing industry, manufacturing newly invented transistors and semiconductors that innovation stations—such as Bell Labs and the company that Bledsoe founded, called Panoramic Research Incorporated—were trying to dream up applications for. Panoramic frequently brought in guests, including Minsky, for conversations about what computers might one day accomplish.

Bledsoe's research team at Panoramic worked primarily on automated pattern recognition and machine reading, giving a computer a whole bunch of data in the hope that the computer could sort through it in a meaningful way. But it was early days, so it was challenging to write programs that actually worked. A magazine publisher asked Panoramic for a computer program that could scan its subscribers' names and addresses off metal-stamping plates so that the company could have a list of its customers. Panoramic created a prototype that worked reasonably well in the lab but failed utterly in the field, where the metal plates were not pristine; they were routinely re-stamped with a new subscriber's address when an old one canceled, and that confused the computer. Panoramic didn't get the contract.

Many of Panoramic's other ideas were moon shots. Without practical products to sell, the company might have run out of money, but luckily it had a government patron: the Central Intelligence Agency (CIA), which was quietly channeling money to scientific efforts that might assist its intelligence gathering. The agency appears to have been interested in a line of research Panoramic was pursuing, an attempt to build "a simplified face recognition machine."

Bledsoe warned that it wouldn't be easy. There were "millions or even billions" of possible variations of a face given head rotation and tilt, the lighting of a photograph, facial expression, aging, and hair growth. But he thought something like the Bertillon system might work: placing dots on key locations of the face and then developing an algorithm that would account for how the distances between the dots would change as the head turned. That might require, Bledsoe admitted, "a certain amount of cooperation between man and machine." A person would need to draw the dots, specifically Bledsoe's teenage son and a friend of his.

Bledsoe listed a handful of reference texts that had informed his thinking. Computer science papers made the list, as did Alphonse Bertillon's textbook for taking measurements of prisoners. A more surprising inclusion was Cesare Lombroso's *Criminal Man,* the 1876 treatise that had described lawless individuals as a less evolved species whose faces bore the telltale features of "primitive races." Bledsoe's proposal appears to have been compelling enough to get the agency's support. In 1965, the CIA funded "the development of a man-machine system for facial recognition," according to a heavily redacted memo on its research and development spending that year.

The teens marked up seventy-two photos of eight men, and then applied millimeter rulers to them, taking twenty-two measurements of each face. Bledsoe's colleague, Helen Chan Wolf, who would later help develop the world's first autonomous robot, wrote an algorithm to crunch all those measurements and create a persistent identifier that should be able to match a man to his measurements no matter which way his head was turned. This system, which was far from entirely automated, worked at matching the eight men to their own photos, and the researchers considered it a success, if on a small scale. "The important question is: How well would this system do on a very large file of photographs?" they wrote.

Beyond creating ample employment opportunities for teens, it probably wouldn't have worked very well. But what if it had? If those technologists had perfected facial recognition technology and put it into the hands of American intelligence officers who still held sacred the ideals of the conservative 1950s, what would the 1960s have been like?

In Alabama, authorities had just tried to force the NAACP, a racial justice organization the state wanted to ban, to turn over a list of all its members—a request that the Supreme Court ultimately deemed unconstitutional. But with facial recognition technology, the state wouldn't have had to ask for that list of people that it wanted to monitor or intimidate; it could have just photographed attendees at NAACP meetings and protests and then matched them to their photos on government documents.

In 1963, the year Woody Bledsoe proposed his facial recognition work, Martin Luther King, Jr., gave his famous "I Have a Dream" speech on the steps of the Lincoln Memorial in Washington, D.C. The reverend turned political activist decried the plight of Black Americans, who were "sadly crippled by the manacles of segregation and the chains of discrimination," and called for the country to make good on its promise that all men are created equal. FBI agents wiretapped King's home and offices and bugged his hotel rooms, eventually recording evidence of his extramarital affairs. They thought they could silence him with blackmail. They made a tape to send to King and his wife, along with an anonymous letter that attempted to shame him into committing suicide. Those officers used the most powerful surveillance technologies of their era, bugs and wiretaps, to stymie King and his work. What would they have done with more powerful modern technologies for mass surveillance?

Would young men have been able to dodge the draft during the Vietnam War, or would they have been tracked down with real-time cameras and been conscripted? Would war objectors have been able to protest without fear of being identified and fired?

Same-sex relationships were stigmatized in the 1960s. Would it have become too dangerous to socialize publicly or to go to gay bars for fear of a bigot with a facial recognition app outing you?

Anonymity provides powerful protection for those who don't conform to the status quo. The availability of perfect surveillance in the early 1960s might have preempted the meaningful social change to come. Luckily, it took much longer than that to get the technology anywhere near working at scale.

IN THE DECADES after Woody Bledsoe's work with the CIA, researchers around the world tried to replicate his methods using incrementally more powerful computers. The most significant leap forward came in the 1970s, when a Japanese researcher named Takeo Kanade got his hands on what was then a computer science bonanza: eight hundred photos of human faces.

The photos had been taken at the 1970 World Exposition in Osaka. Though the fair's theme had been "Progress and Harmony for Mankind," the reigning global superpowers couldn't help but try to one-up each other. At dueling pavilions the Soviets had displayed a replica of the spacecraft that had launched the first astronaut into orbit, while the Americans had showed off the two-pound rock they had brought back from the first manned trip to the moon.

Another popular exhibit, attracting thousands of visitors, was a Japanese display called "A Computer Physiognomist's Magnifying Glass." Visitors were told that a computer could determine, based on how they looked, which celebrity personality they had. It was AI Aristotle.

This juxtaposition of the ancient "science" of physiognomy with newfangled tech had been brought to the World's Fair by the Nippon

Electric Company, or NEC.* Visitors to the exhibit sat down in front of a TV camera for about ten seconds so that a digitized photo could be created, converting their faces into a series of pixels, each one assigned a value from very dark to very light. Then the program extracted the major contours of their features, based on what the pixel patterns indicated was an eye or a nose or a mouth. Those key points were then compared to the pointillistic faces of seven famous people, including Marilyn Monroe, Winston Churchill, and John F. Kennedy, to declare the visitor's personality type: "You're a Marilyn!"

"Though the program was not very reliable, the attraction itself was very successful," Kanade wrote years later. At the time, he had been getting his PhD in electronics engineering at Kyoto University. His adviser had helped NEC come up with the idea for the "Magnifying Glass" exhibit and had been allowed to keep many of the photos taken. They were stored on reels of magnetic tape the size of dinner plates, and when Kanade was looking for an idea for his thesis project, his professor handed them over to him. "If you can process all these pictures, that alone will constitute your PhD," his adviser told him.

There were young faces and old, men and women, some with glasses, some with facial hair. It was an incredible variety. When Kanade looked at the NEC program's analysis of the faces, he had to laugh. It had labeled eyebrows as noses and the side of someone's mouth as an eye. He got to work improving on the program so that it could more accurately extract dozens of important facial reference points: the edges of the face, the exact center of eyes and nostrils, the top and bottom of lips, and so on.

Kanade's program was confused by mustaches and glasses, but when Kanade tried it out on seventeen men and three women who didn't have facial hair or eyewear, it performed reasonably well. It was able to match two different photos of the same person in fifteen of twenty cases, prompting Kanade to optimistically declare in a paper that a computer could extract facial features for identification "almost as reliably as a person."

* NEC is now one of the leading sellers of facial recognition algorithms.

That was an overstatement. But it *was* the first fully automated form of facial recognition, with no teens involved, and researchers around the world took notice. For that and other contributions, Kanade became widely recognized as a pioneer in the field of computer vision and was recruited by Carnegie Mellon University in the United States.

Other computer scientists followed Kanade's example, but their experiments succeeded only in matching the faces of a small sampling of individuals. That wasn't going to help spot a wanted criminal or dangerous person in a large crowd. With no practical applications, interest faded and progress stalled.

AND THEN ALONG came Matthew Turk. In 1986, Turk was a junior engineer with the defense contractor Martin Marietta, where he was working on one of the world's first self-driving cars. Called the Autonomous Land Vehicle but nicknamed "Alvin" by Turk's team, it was the size of an ice cream truck with eight wheels, a camera on its roof, a laser scanner where its windshield would have been, and a belly full of computing equipment.

The vehicle was white with blue stripes; the acronym "ALV" was painted on its side in a large, stylized font. Where the driver's door would normally have been, there were logos: one for Martin Marietta (which later merged with Lockheed to become Lockheed Martin) and another for DARPA, which had paid at least $17 million for the hulking vehicle as part of a billion-dollar Strategic Computing Initiative intended to create intelligent machines capable of "far-ranging reconnaissance and attack missions."

DARPA stands for Defense Advanced Research Projects Agency, America's fear of being bested on the world stage by the Soviet Union made manifest. President Dwight D. Eisenhower signed legislation forming DARPA in 1958, shortly after the Soviets pulled ahead in the space race with Sputnik, the first satellite to orbit the planet. Since then, DARPA had tapped what seemed to be a bottomless pit of money for research-and-development projects to ensure America's technological dominance in the military and beyond. By 1986, when Alvin came along, DARPA had already funded projects that would evolve into the internet, automatic target recognition, and the Global Positioning System, or GPS.

On a summer afternoon, journalists were invited to the outskirts of Denver, Colorado, to see what Alvin could do. In a preview of its limitations, the journalists were warned not to get mud on the track. Alvin was driving itself by taking depth readings with its laser and analyzing the road markings coming in from the camera on its roof, an impressive feat, but its programming allowed it to recognize only a perfect version of the world around it. Mud and even shadows were insurmountable obstacles.

Despite being one of the most expensive cars ever created, Alvin did not impress its audience. "The current top speed of the high-dollar coach is only 6 mph," grumped one journalist. Despite the careful pace and clean track, Alvin steered itself off the road. According to one history of DARPA, the military was just as disappointed as the journalists. One officer "complained that the vehicle was militarily useless: huge, slow, and painted white, it would be too easy a target on the battlefield." He obviously didn't understand that Alvin was meant to be a proof of concept, not a *Star Wars* battle robot ready to start taking out Ewoks.

DARPA's goal was grander and more subtle than building a war machine the army could use right away. The agency hoped to seed the research—and the researchers—necessary for remarkable, and ongoing, technological breakthroughs. The point of the billion-dollar investment wasn't just to make an Alvin; it was to groom engineers like Matthew Turk.

Drawn to the challenge of improving computer vision, Turk applied to the MIT Media Lab, a new technology department whose well-branded mission was to "invent the future" and to make real the stuff of science fiction. When Turk arrived at the Massachusetts Institute of Technology in 1987, with shaggy brown hair and a handlebar mustache, Marvin Minsky was still there, even more gnomish than he had been in his youth, the fringe of hair remaining around his head grayed. His main accomplishment to date was a robotic arm with a camera that could "look" at a stack of children's blocks, take it apart, and then build it again; the culmination of three decades of his AI work was literal child's play.

Turk's PhD adviser, a computer scientist named Alex "Sandy" Pentland who had also worked on the Autonomous Land Vehicle, suggested that Turk pursue automated facial recognition despite its being a challenge that Minsky had just deemed practically impossible. "Face recognition was thought of as being one of the most sophisticated capacities of humans, with multiple levels of representation and reasoning," Pentland said. It was "a classic example of human intelligence" that seemed beyond the reach of computers at that time.

But face recognition had two things going for it: a new computer technique out of Brown University for compressing an image of a face and, perhaps even more important, corporate funding. A Maryland-based radio and TV ratings company called Arbitron was willing to bankroll the project to gain an edge on its competitor, the industry heavyweight Nielsen. Nielsen had equipment installed in the homes of thousands of families, tracking what they watched on television. But the equipment could track only that the TV was on and what channel it was tuned to, not who was in the room. In order to gather more precise ratings and better-targeted advertising, this was sometimes paired with a "people meter," where members of the household could push a button when they sat down to watch TV and then again when they left. But people weren't reliable button-pushers, so Arbitron wanted an "automatic people meter"—facial recognition that could keep track of who was in the room and when.

"So that was really my first order of motivation," Turk said. "A box sitting on top of the TV, watching the family, to log people in and out basically." The family watches the TV, and the TV watches them back.

The Media Lab had embraced the approach to AI that Minsky had rejected: using the natural world to inform the building of machines. Technology students were encouraged to interact with other disciplines, such as the neuroscience department, because the inner workings of the human mind might inform how they programmed computers. When Turk looked at the research into biological vision and how the brain processes faces, he discovered that scientists were still trying to figure it out.

The human ability to remember thousands of faces is remarkable, one that develops so early that a baby, potentially within hours of birth, can distinguish its mother's face from others in a room. Most people can still recognize a person who shaves off a beard or puts on glasses, but for computers at the time, that was inconceivable. Neuroscientists thought that a special part of the human brain was devoted solely to processing faces. (This is still a matter of debate, one cognitive neuroscientist has said, as this fusiform part of the brain also lights up when an ornithologist looks at a bird or a car connoisseur looks at a vehicle.)

Human facial recognition seemed to be holistic, not just a sum of the parts. People couldn't easily say which nose belonged to whom when looking at the schnozz alone. Test subjects could recognize someone from just half of their face but struggled when that half was paired with half of someone else's. People could easily match a house to a version of it that was flipped upside down but were slow to do the same with a face. Some facial features mattered more than others: Covering the upper parts of the face made someone harder to recognize than masking the mouth and jaw (as many a superhero knows). Intriguingly, some people excelled at facial recognition—this appeared to be partly genetic—while some were truly terrible at it.

When it came to automated face recognition, Turk believed that progress had leveled off because researchers were so focused on dots and measurements. That's not how humans do it. We aren't walking around with rulers, holding them up to our colleagues' noses in order to remember their names. Rather than forcing the computer to use a ruler, Turk thought he should give the computer a bunch of faces and let it decide the best way to tell them apart.

Turk borrowed a technique from researchers at Brown University who had developed a method to digitally represent a face while minimizing the amount of computer memory it took up. They had created a facial composite of 115 Brown undergrads, and then stored each one as a mathematical calculation of how it differed from the average face. This worked in great part because all the undergrads were "smooth-skinned caucasian males," with their main deviations from one another occurring in the eyes, the eyebrows, the jaw, and the area around the nose. The result for each individual was a caricature rather than a high-fidelity profile photo, but it was an interesting idea, and, as it would turn out, very useful for someone who was trying to build a facial recognition system.

"Any face is a combination of a bunch of other faces," Turk said, explaining what the Brown researchers had shown. It was a kind of recipe; rather than combining flour and sugar, it was a matter of how much "weight," in engineering terms, to assign to each ingredient—a heavy helping of thick brows, a dash of high cheekbones, a sprinkle of dimples. Instead of trying to match one facial map to another, Turk's system would allow a computer to analyze visual information in a way only it could—in bulk—taking into account not just how human faces are different but how they are fundamentally the same.

Turk gathered sixteen volunteers from the Media Lab, again all men and mostly white, and asked them to tilt their heads this way and that as he took their digital portraits using a hefty, twenty-pound video camera hooked up to a computer. He replicated what the Brown researchers had done to derive facial ingredients that, when combined the right way, could represent each man: seven gray, ghostly-looking impressions that he called *eigenfaces,* invoking a German mathematical term for directional stretching used by the Brown team.

To do a real-time face match, Turk rigged up a video camera, an image-processing system from a company called Datacube, and a desktop computer to run his software. When the system sensed movement in the room, it looked for a face, converted it into a mix of *eigenfaces,* and then compared that mix to the sixteen stored in its library. It would either register a match, tag someone as unknown, or determine that the thing it saw moving was not actually a face. Arbitron never wound up using the system, but Turk wrote a paper about what he had done, and it was impressive enough that he was invited to present it at a major academic conference in Hawaii.

In the summer of 1991, Turk boarded a flight to Maui on MIT's dime. The conference was at the Kāʻanapali Beach Hotel, an oceanside resort across the street from a lush golf course. He was thrilled, and not just because his fiancée had joined him. The most prominent computer vision researchers in the world were there, and several were in the room when he gave his presentation, using an overhead projector to show off his ghoulish *eigenfaces* on the wall behind him. After he finished, a man in the audience with a big, bushy white beard stood up with a question. Turk recognized him immediately. His mathematics textbook was used in engineering graduate courses.

"This is old hat," Turk remembers him saying. *Eigenfaces* was just a sexy new name for an old technique in engineering called *principal components analysis,* a statistical breakdown of a bunch of data. It was the math equivalent of reading the CliffsNotes for *Moby-Dick*. People had tried it before in computer vision, he said, and it doesn't work. Turk was intimidated but managed to muddle his way through a response. Teaching a computer to recognize just sixteen people was almost laughable, but it *had* mostly worked and it pointed to a new way forward.

Relieved the presentation was over, Turk headed back to the pool, feeling glad that, even if his work hadn't wowed a leader in his field, it had gotten him a trip to Hawaii with his future wife. The morning after a special awards dinner banquet, which Turk had skipped because it cost extra, he arrived late to a talk. The person sitting next to him leaned over and whispered, "Congratulations."

"For what?" Turk asked.

"You won a 'best paper' award."

IN THE YEARS to come, Turk's paper took on a life of its own. It was taught at universities and cited more than twenty thousand times by other researchers. "I had no idea it was going to become this monster that it turned out to be," Turk said.

Turk's *eigenface* approach represented a turning point, a moment that suggested that computers could data crunch their way to "seeing" the world and people around them. With enough data, enough photos of a person, a computer could start to understand what a person looked like from different angles, with different expressions, and even at different ages. The spark that computers needed to go from seeing a collection of facial features to seeing an individual was *a lot* of faces, a bit more freedom in determining how to analyze them, and more computing power.

Facial recognition became a puzzle that looked solvable. Researchers were motivated not just by funding that began to flow in from the federal government but by "technical sweetness," a term used by the science philosopher Heather Douglas to describe the delight that scientists and engineers feel when they push innovations forward, which may overpower any concerns they should feel about that progress. The concept was most famously expressed by physicist J. Robert Oppenheimer after he helped develop the atomic bomb: "When you see something that is technically sweet, you go ahead and do it and you argue about what to do about it only after you have had your technical success." After the first successful test of an atomic explosion, Oppenheimer said a quote from the Bhagavad Gita had flashed through his mind: "I am become death, the destroyer of worlds." That, said Douglas, was the bitter aftertaste of technical sweetness.

Shortly before Turk got his PhD in 1991 and largely left facial recognition behind, a documentary film crew came to the MIT Media Lab to interview him. The documentarians wanted a demo, so he recruited a few students and researchers—one woman and three men—to play the part of a family living in one of Arbitron's tracked households. When the prototype "automated people meter" successfully detected a person sitting on the couch, their name popped up above their head in a sans-serif white font on Turk's desktop computer. It was working perfectly until the film crew surprised Turk by bringing in a black Labrador.

The system tagged the dog as "Stanzi," the one woman in the experiment. Looking back on the documentary decades later, Turk said it had happened because the researchers had set the bar low for detecting a face and never tested it with an animal in the room. But it was a revealing moment in early computer vision history. In the experiments at Brown and then at MIT, the "average face" shown to the computer was consistently that of a white man. Turk's program was trying to identify someone by how much he deviated from the "average face," and apparently, the faces of both a woman and a dog fell into the range of "major deviation." In the decades to come, even as AI systems became far more advanced, they would struggle with the faces of women and non-white people, even continuing to mistake them for animals.

Turk said that the big issues that would come to plague facial recognition technology—privacy and bias—had not been on his mind thirty years earlier. The technology was so rudimentary that Turk couldn't conceive of its ever having a significant influence on people's day-to-day lives. In the documentary, though, Turk does seem to be at least a little concerned. After another expert called the idea of facial recognition cameras in bedroom TVs "Orwellian," Turk defended the system, saying that the cameras weren't recording but simply collecting information about who was in the room. "It's rather benign," he said.

The interviewer pressed him on that: "So would people be willing to have one of these in their bathroom or bedroom?"

"I wouldn't," Turk responded, laughing. "But don't put that on your show."

CHAPTER 5

A DISTURBING PROPOSAL

On election day 2016, Hoan Ton-That and Charles Johnson were at Donald Trump's invitation-only party at the New York Hilton.

"It was a sorry affair," Johnson said. "All the reporters there were like 'What are you going to do when Trump loses?' And I was like 'Uh, he's not going to lose.' And they were like 'You're crazy.'"

Johnson had flown in from California, leaving his pregnant wife behind. He thought that Trump might just eke out a win on the sheer power of his celebrity. Trump's reality TV show, *The Apprentice,* had been watched by millions of people, and it had reinforced the real estate mogul's carefully cultivated image as a successful businessman—despite investigative reporting to the contrary.

Looking for something livelier than the resigned crowd at the Hilton, Johnson and Ton-That took an Uber to the Gaslight lounge in the Meatpacking District, where Gavin McInnes, a co-founder of *VICE* magazine, was hosting a gathering of the Proud Boys, later designated a hate group by the Southern Poverty Law Center. McInnes had formed the "drinking and fighting" club the previous year but had only recently declared its existence publicly, writing online that it was for "Western chauvinists who refuse to apologize for creating the modern world." Women were not allowed.

The Proud Boys' election night party had swag: Trump bumper stickers and a basket full of masks, including the face of conspiracy theorist Alex Jones, in honor of Hillary Clinton's remark that half of Trump's supporters could be put into a "basket of deplorables." They were "racist, sexist, homophobic, xenophobic, Islamophobic," she said, and Trump had "lifted them up." The Trump movement proudly embraced the label.

Under the crimson light of the Gaslight lounge, a man wearing a full-face metal mask was holding court. He was dressed as Lord Humungus, a villain from the apocalyptic *Mad Max* movies, his large body encased in a black leather bondage outfit, spiked straps across his naked chest. "The Proud Boys will not be defied!" he said in a theatrically deep voice, pumping the crowd up.

Gavin McInnes, dressed incongruously in a Hawaiian shirt, was reveling in his role as the leader of a resurgent right-wing movement. He gave a speech rhapsodizing about the way a Trump victory would empower the Deplorables. "If Trump wins . . . well . . . I don't know how to talk right now," he said into a microphone. "That's like 'What should we do if we cum?' " He rounded his mouth and closed his eyes as he faked an orgasm.

"Light a cigarette!" yelled someone who sounded like Charles Johnson from the front of the audience. When McInnes finished speaking, the assembled crowd began shouting, "USA! USA! USA!"

It all made for a bizarre spectacle, the line between performative villainy and authentic aggrievement impossible to draw. It was raucous and grew more so as, in a stunning turn, the election results trickling in started to favor the Proud Boys' candidate. That was when Johnson and Ton-That decided to leave. "There are too many dudes here," Johnson recalled saying to Ton-That. "This is uncomfortable."

With the gender ratio topping the list of possible discomforts, they rejoined the crowd at the Hilton, which had become livelier as it seemed increasingly possible that Trump might prove the election pundits wrong. Johnson snapped photos of the conservative luminaries in attendance and posed for a few with Ton-That, a row of American flags behind them on the stage. Ton-That wore a mixed-print shirt and Johnson sported a MAGA hat and a pair of over-ear headphones around his neck. The room had almost no Trump paraphernalia decorating its walls, as if the organizers hadn't expected that it would be a night they'd want to memorialize.

Johnson chatted with Erik Prince, a former Navy SEAL whose private military force, Blackwater, had been paid hundreds of millions of dollars by the United States government. After Blackwater mercenaries had committed atrocities in Iraq, murdering innocent civilians, including a nine-year-old boy, Prince had stepped down as CEO and sold the firm. At the Hilton, Johnson snapped a photo of Prince smiling joyously, standing with his wife, who was making the OK hand gesture. The excitement in the room was growing. The results were looking better and better for Trump, who had to huddle with his team upstairs to rewrite his remarks.

Sometime after 2:00 A.M., media outlets began to call the race. Trump had lost the popular vote but won the electoral college. Around 3:00 A.M., after getting a call from Clinton congratulating him, Trump delivered his victory speech. The candidate that the room had been rooting for, against all hope, would become the forty-fifth president of the United States.

"Hoan and I are hugging and we're so happy," Johnson said. In the wee hours, after Trump's speech, Ton-That and Johnson went to a nearby diner with a big group of celebrants, including Prince, who paid the tab. "It's very clear the world is going to change with Trump's election, but it's not clear how it's going to change in our own lives yet," Johnson said.

Johnson saw it as a victory over staid establishment politics and as an opportunity: He felt that the new political administration would be kind to people like him, Ton-That, and Schwartz. A week later, on November 15, 2016, they registered a new website, smartcheckr.com. It would be a tool that could theoretically identify and root out extreme liberals.

BY EARLY JANUARY 2017, Ton-That, Schwartz, and Johnson had drafted an initial business agreement, listing the three of them as co-founders with equal equity in their new company, SmartCheckr LLC. Schwartz and Johnson were supposed to provide the capital while Ton-That built the technology: a search app for people to turn up their political leanings, social media footprints, and possible vulnerabilities, such as passwords exposed in a data breach.

Johnson boasted that he had powerful connections that would help the company. He was spotted that month on Capitol Hill, chatting with congressmen. *Forbes* reported that he was working with members of the executive transition team, including Trump donor Peter Thiel, to make recommendations for high-level cabinet positions. At the end of January, Johnson did a celebratory "Ask Me Anything" session on the online forum Reddit. "I essentially had a choice between being in the Trump administration or working to fill it with friends and allies," he claimed. He said he hoped that Trump would "crush the communists in Silicon Valley" and, when asked about his thoughts on the Holocaust, volunteered that he did not believe the "six million figure," saying he believed that far fewer people had actually been killed in concentration camps. "You can't really discuss any of this stuff without being called a Holocaust denier which I am not," he wrote, correctly predicting a label that would be attached to his name for years to come.

SmartCheckr LLC wasn't officially registered in New York until the next month, but Ton-That already had a prototype—a user could enter someone's email address and it would surface that person's photo and their social media accounts by pulling information from Facebook, Twitter, LinkedIn, and Pinterest. But the app was more than just a specialized Google; it included an analysis of whom people followed, providing insight into their political orientation.

This time in SmartCheckr's history is shrouded in secrecy, but it appears that one of the app's first uses was to vet attendees for the DeploraBall, a celebration on the eve of Trump's inauguration. The event was organized by a loose collective of conservatives, including Ton-That and Johnson's dinner companion Mike Cernovich and a Peter Thiel associate named Jeff Giesea. A person affiliated with the event, who asked not to be identified, confirmed that Ton-That had reached out to the organizers, offering to provide a tool to check all the attendees and find out whether they were "alt-right or Antifa," a derogatory term, with the emphasis on the second syllable, for antifascists and others on the far left.

The DeploraBall was to be held on January 19, 2017, at the National Press Club, a venue a few blocks away from the White House, and would feature an open bar, light hors d'oeuvres, and "great people of all backgrounds." "Our goal is to celebrate and honor people like YOU—the passionate citizens who worked social media, knocked on doors, and endured harassment to support our President-Elect," the Eventbrite page said. "You deserve this."

The "unofficial inaugural ball" was a gathering place for the online cohort who had bonded over their allegiance to the polarizing presidential candidate. The world had laughed at them, scorned them, called them racists, yet they had remained steadfast, putting LOCK HER UP signs on their front lawns, turning out voters in the swing states, and proudly wearing MAGA hats to the grocery store. They had persevered, and now Trump was taking over the Oval Office.

"The internet won," Johnson said, referring to a very particular right-wing slice of the internet. Outside that slice, many Americans were devastated by Trump's win, and vocally so, both online and off. Liberals were horrified that a former reality TV star who had been openly misogynistic, xenophobic, and bigoted during his campaign was going to be in charge of the country for the next four years.

FUCK TRUMP was graffitied on walls across the country after the election. Protesters chanting "Not my president" walked the streets of dozens of cities. In Washington, D.C., a group called the Antifascist Coalition wanted to make clear to the world that they did not welcome Trump and the right-wing circus he was bringing to the nation's capital. One of the coalition's organizers, a longtime antiauthoritarian anarchist named Lacy MacAuley, who had helped organize D.C.'s Occupy Wall Street offshoot a few years earlier, was anguished by Trump's election. She had a "sweeping fear," as she described it, "that people of color in our country were going to be persecuted and suffer."

She and her collaborators decided to organize a protest outside the DeploraBall but also to send some members inside, undercover. They wanted to document anything along the lines of what *Atlantic* reporters had captured at an alt-right think tank in the days after the election, when a speaker invited the audience to "hail Trump," leading some attendees to stand and give Nazi-style salutes. Slight, with long blond hair, MacAuley planned to get a cheap ticket and attend in disguise as a "right wing woman" by doing a "ferocious amount of grooming"—fancy hairstyle, heavy makeup, and an elaborate dress. "I thought I could code switch and look just like another white lady," she said. She registered for the event, but for some reason was never able to get a ticket.

As the antifascists made plans for a smorgasbord of inaugural protests, more and more people wanted to join them. That wasn't strange; the country was fractured and fuming. But a couple of the newbies struck them as a little off, such as a quiet guy named Tyler. Tyler had a vague backstory and didn't seem well educated about liberal causes, making MacAuley and others suspect that he might be a conservative operative. Had the infiltrators been infiltrated, as in a *Spy vs. Spy* cartoon?

MacAuley said that the D.C. Antifascist Coalition had hatched a harebrained scheme to find out. A few members of the coalition invited Tyler to Comet Ping Pong—selected humorously, she said, because it was the unfortunate pizzeria at the center of the #PizzaGate conspiracy theory. Over beers there, the antifascists told Tyler about a plan to stink bomb DeploraBall attendees and set off the sprinklers inside the National Press Club to disrupt the proceedings. The "plot," which seemed ripped from a dumb 1980s movie, had been made up, MacAuley said, in the hope of feeding false information to their enemies and outing Tyler if he was a spy.

The plan backfired terribly. They were right: Tyler *was* an infiltrator, employed by a conservative investigative media group called Project Veritas, which specialized in secretly recording people. Tyler had been filming all the meetings he had attended, including the one at the pizzeria. Project Veritas published an online report about the planned hijinks and notified authorities, who took the "bomb threat" seriously. The men who had confided in Tyler, even if jokingly, were arrested, charged with conspiracy, and ultimately pled guilty to misdemeanors.

The DeploraBall, on the other hand, went off without a hitch. MacAuley still showed up that night, but only to protest outside the building. There were no stink bombs and no Nazi salutes. There were men in tuxes and MAGA hats, women in party dresses. Gavin McInnes and Peter Thiel were among the attendees. Thiel stayed for just thirty minutes.

SMARTCHECKR TOOK CREDIT for blocking unwanted guests at the DeploraBall in the oddest of places: a product pitch it made to the

Hungarian government, one of the first customers the fledgling company approached.*

Hungary's far-right prime minister, Viktor Orbán, described as a "strongman" by the press, was at the head of a growing authoritarian, populist movement across Europe, much like the one Trump had ridden to power in the United States. The American Right admired Orbán as a leader who was beating back progressives and winning the culture wars in his country: He had banned transgender people from legally changing their sex and built a barbed-wire border wall to keep out immigrants. Trump adviser Steve Bannon would later call Orbán "Trump before Trump." Ton-That, Johnson, and Schwartz wanted to offer him their services.

The Smartcheckr team had been busy in the months since Trump took power. Three months after the inauguration, in early April 2017, Hoan Ton-That reached out to a man named Eugene Megyesy, an adviser to Orbán whose contact information Charles Johnson had in his right-wing Rolodex.

"Hi Gene," wrote Ton-That in his email, with Johnson and Schwartz cc'd. "Attached is a summary overview of our background-search system, Smartcheckr, and how, we believe, it can help support Hungary in its border-security and enforcement efforts." Attached to the email was a PowerPoint presentation entitled "Smartcheckr: The Future of Border Security."

It described Smartcheckr as a "proprietary, first-of-its-kind search technology that can create instant profiles of individuals based on email addresses and/or headshots." Smartcheckr could find out whether a person was involved "with guns, criminal activities, radical political or religious activities or other adverse pursuits." The DeploraBall, it said, was a successful deployment of its "Filter for Extremist Politics at Large-Scale Event." Smartcheckr had evaluated more than 550 guests in "less than 5 minutes" and identified those who were "high risk." One of them was Lacy MacAuley.

The presentation included what they had dug up on MacAuley

*They made it "SmartCheckr" on their corporate registration but eventually started styling it as Smartcheckr.

based on the email address she had used to register for the event: three photos of her; nine of her social media accounts; two "hacked databases" where her passwords had shown up (Forbes and LinkedIn, which had both suffered from recent data breaches); and an analysis of her Twitter activity with a "WARNING" that she was following "10 antifa account(s)." One of these so-called extremist accounts belonged to Showing Up for Racial Justice, a national network of white people who opposed white supremacy.

The PowerPoint presentation for Hungary said that Smartcheckr had been fine-tuned to identify people that Orbán considered enemies: pro-democracy activists who believed in "open borders." Smartcheckr's "research team" had combed social media, including likes and follows, for "virtually every individual affiliated with the Open Society Foundations and other related groups," the pitch to Orbán's adviser said. "This enables Smartcheckr to instantly flag any individual who works for or is connected in any way to these organizations, their officers or employees."

The Open Society Foundations provide grants to civil society groups around the world to advance "justice, democratic governance, and human rights." The organization was founded by the billionaire George Soros, who was from Hungary; of Jewish descent, he had survived the Nazi occupation there. Orbán despised Soros's group, accusing it of undermining Hungary's sovereignty, and wanted it out of his country.

Smartcheckr's background screening tool could assist, the company said, by scanning every passenger on incoming flights to Hungary to "flag problem individuals" and by patrolling the border using "drone technology," with the caveat that it was "in development." The company had established, it said, a "unique joint venture with Oculus Virtual Reality founder Palmer Luckey." Luckey was a tech wunderkind who had sold his virtual reality startup to Facebook, but then was ousted from the company for funding a meme-maker who had spread pro-Trump propaganda during the 2016 election. Luckey hadn't announced it publicly yet, but he was working on a new company that aimed to create sensor-based border walls. Smartcheckr claimed to be working with the military startup on camera-equipped drones "technologically integrated with the Smartcheckr search platform" that would enable "instant background reviews of individuals attempting to cross the border."

These surreal claims were far beyond what Smartcheckr was actually capable of at that time, judging from internal emails. Ton-That had coded a tool that could provide a person's race and gender based on a photo, but not one that could produce someone's name based on a flyby picture. What Smartcheckr actually had at that point wasn't much more than what the Hungarian government could have gotten by googling someone or looking them up on one of the various obnoxious people-search tools that haunt the internet with names such as BeenVerified, TruthFinder, and Instant Checkmate. But that is what startups do: fake it till they make it.

The presentation did contain a clue about a method Smartcheckr would need to employ to get information from social media sites when they stopped letting people like Ton-That use their built-in developer tools: unauthorized scraping. "Scraping" sounds kind of nasty, maybe even a little painful, but when it comes to the internet, it's just the act of downloading data en masse; it's been going on since the World Wide Web was brought into existence in 1989. It has long annoyed the sites being scraped; early webmasters complained that their site servers were taxed and slowed by the incessant bot visitors. To deter scrapers, companies occasionally sue them, alleging "theft" of what they consider proprietary material—even though it's publicly available on the internet.

Sites such as Facebook and LinkedIn prohibit third parties from scraping their content and have technical measures in place to block them, most notably throttling mechanisms called data and rate limiting, which cap how much a single user can get, and how often they can get it. But the presentation to Hungary described a way to circumvent those protective measures. It said that Smartcheckr "gathers social media data through an undetectable method that accesses more than 20k residential IP addresses worldwide." This would disguise the bots and make them look like human users sitting at lots of different homes. This "undetectable method" sounds similar to a commercial service called a *proxy network* that almost any business can buy.

The outlandish PowerPoint presentation that Smartcheckr whipped up for the Hungarians, using a default template with a purple palette, looked clownish and led nowhere. It was telling that the team had no compunction about selling its tech to an authoritarian state intent on silencing its critics, but it was not unprecedented. By 2017, government authorities and law enforcement had been getting pitches for subpar facial recognition technology for decades.

CHAPTER 6

THE SNOOPER BOWL (2001)

On January 28, 2001, the New York Giants faced off against the Baltimore Ravens in the Super Bowl. The big game took place in Tampa, a party-friendly city on the west coast of Florida, in a stadium that seated more than seventy thousand people. The halftime show featured the grizzled rockers of Aerosmith, the R&B singer Mary J. Blige, the fresh-faced boy band 'N Sync, the pop phenom Britney Spears, and the rapper Nelly, a lineup engineered to please a wide range of music lovers.

On the Monday after the game, while sports obsessives dissected the Ravens' victory and cultural analysts declared the halftime show one of the best ever, an obscure company based in Massachusetts put out a typo-ridden press release titled "Criminals No Longer Another Face in the Tampa Stadium Crowd."

Apparently several technology companies, including the defense contractor Raytheon and a facial recognition vendor named Viisage Technology, had partnered with the local police "to monitor potential criminal activities." Unbeknownst to those in the crowd, they had deployed a "surveillance and facial recognition system" in Tampa's stadium because "not everyone comes to sporting events with good intentions," as they put it in the press release.

Viisage, a company that created photo management systems for DMVs and corporate ID badges, had taken the *eigenface* method pioneered by Matthew Turk at the MIT Media Lab and commercialized it. During the Super Bowl, cameras at the stadium had captured the faces of the thousands of attendees. Then the system, according to the press release, "continuously compared faces in the incoming crowd to an extensive, customized database of known felons, terrorists, and con artists provided by multiple local, state and federal agencies." Were any of these dangerous characters to be identified, officers "could be dispatched immediately to make possible arrests," the press release claimed, "quickly and discreetly."

It was a first. Facial recognition had never been used on such a huge crowd in the wild* before, at least not that anyone knew about. The *St. Petersburg Times* broke the story on its front page: "Cameras Scanned Fans for Criminals." The New York *Daily News* declared it the "SNOOPER BOWL."

The Tampa police told the *Los Angeles Times* that the Super Surveillance had been a "test" as part of a free trial. The police had mainly been on the lookout for ticket scalpers and pickpockets; the cameras had supposedly identified nineteen people with criminal histories, "none of them 'significant,'" according to the police, yet none of them were stopped to have their identities confirmed, and no one had been arrested, despite the press release's bold claims. Regardless, the department was pleased with the trial.

"It's just another high-tech tool that is available," a police spokesman said. "And yes, we did like it. Very much so."

A noted security expert felt differently. "Oh my God, it's yet another nail in the coffin of personal liberty," Bruce Schneier, a technologist who wrote frequently about security issues, told the *Los Angeles Times*. "It's another manifestation of a surveillance society, which says we're going to watch you all the time just in case you might do something wrong."

A constitutional law professor was "troubled" by the episode but said it was perfectly legal. "People have no reasonable expectation that when out in public, they cannot be photographed," he said.

But this was more than just taking a photo. This system was checking the faces of more than seventy thousand people against the faces of

* "In the wild" is a phrase used often when talking about face recognition, to differentiate between faces caught in perfect conditions, such as a portrait session, from those of people who were unaware their photo was being taken or weren't posing for it.

seventeen hundred wanted criminals, based solely on the hunch that a ne'er-do-well might have been tempted to attend the big game. *The New York Times*'s editorial board came out against "Super Bowl Snooping," saying that citizens were willing to accept security cameras at an ATM but not random surveillance of a huge public event, especially with the risk that an innocent person might be misidentified as a criminal. Viisage's then president, Tom Colatosti, wasn't bothered by the scathing coverage. "There's no such thing as bad publicity," he later said. "We got so many inquiries. The marketing was terrific."

Still, a plan to use facial recognition on the crowds at the upcoming Olympics in Salt Lake City, Utah, was called off. The American Civil Liberties Union (ACLU) of Florida called for hearings, with its director saying the technology's use on an unsuspecting crowd was how "totalitarian societies" operated.

Dick Armey, a conservative congressman from Texas and one of the most powerful lawmakers in the country, issued a rare joint statement with the ACLU, calling on all levels of the government to stop using facial recognition "before privacy in America is so diminished that it becomes nothing more than a fond memory." The technology "should not be used to create a 'virtual line up' of Americans who are not suspected of having done anything wrong," they said.

Armey asked the General Accounting Office (GAO), an investigative arm of Congress, to look into how much money the federal government was spending on facial recognition and who was using it. The press started digging and found that casinos—including three owned by Donald Trump in Atlantic City—had been early adopters. "In the first week after we installed it, we arrested a big group of cheaters," Trump Marina Casino's surveillance director told the *Los Angeles Times*. But a facial recognition vendor cast doubt on the claim. The vendor told the paper that his company had tried out facial recognition in a casino and it hadn't worked because of the uneven lighting and the fact that the cameras were pointed mostly at the tops of people's heads.

More articles came out questioning the technology. Did it even work?

When the Tampa Police Department installed facial recognition cameras later that year in a neighborhood called Ybor City, it went with a Viisage competitor with a confusingly similar name, the Visionics Corporation, which also provided the technology for free on a trial basis. Meanwhile, the Pinellas County Sheriff's Office had set up a facial recognition database that could comb through Florida's mug shots. One reporter found that the agencies had done almost no research on how effective the technology was, so she did it for them. She talked to other police departments and government agencies around the country that had tried out facial recognition and they all said the same thing: It wasn't ready for prime time yet.

Joseph Atick, an academic turned entrepreneur who had cofounded Visionics, was worried about the bad press. He thought the publicity stunt at the Super Bowl had been a mistake—a potentially fatal one for the industry—so he had gone on a media blitz, talking about the good the technology could do. He told journalists and government officials that automated facial recognition was an important tool for public safety and reliable enough to be deployed in the real world.

Atick kept a diary during that time, and there he was more candid, writing scathingly about the use of his competitor's system on thousands of football fans: "The reality is the system was a total fraud, it captured not a single face let alone matching against anyone in the database."

His contempt stemmed from the common animosity between competitors, but it also had to do with the *eigenface* technique: "the face as a composite of different possible faces." Atick didn't think the technique worked at scale (and eventually the rest of the industry would come to agree with him). Atick's own algorithm worked by looking at a few key points on a face, a combination of which made each face unique. That was closer to the right approach, but it worked best when a computer decided what the key points were. That would require more data and more powerful computing than anyone had, so it, too, was many years away from feasibility.

But would the industry have that time? The intense public anger and perception that the technology was "evil" worried Atick, as did the pressure coming from Washington, D.C. It seemed possible that facial recognition might be outlawed entirely. "Dick Armey basically vowed to put me out of commission," he recalled years later. "The summer of 2001, I was convinced it was our last summer. We would need to go do something else."

Then the tide turned—dramatically.

ON SEPTEMBER 11, 2001, members of a group called al-Qaeda hijacked four commercial airliners and turned them into missiles aimed at symbols of American power. One plane crashed in a Pennsylvania field after a passenger revolt, but the others reached their destinations: the Pentagon and the Twin Towers of the World Trade Center in New York City.

"Immediately there was a huge current, and we were all carried into it," Atick said. "People wanted technology. People wanted more security. People couldn't care less about privacy."

The attacks on New York City and the Department of Defense quickly transformed the United States into a more fearful nation, where security concerns increasingly took precedence over privacy. After declaring a war on terrorism, President George W. Bush launched military operations in Iraq and Afghanistan. Congress passed the USA PATRIOT Act, giving law enforcement agencies greater access to Americans' financial and communication records so they could more easily track suspected terrorists and their money flows. The act also included a mandate that the government assess the accuracy of biometric tracking technologies, including facial recognition technology. Surveillance company stock prices surged.

Atick and other vendors began talking with airports about scanning travelers' faces to reassure people that advanced technology was being deployed to protect them. Atick knew that the technology wasn't up to the challenge of identifying terrorists in a security line in real time and wouldn't be for another ten or twenty years. But he agreed to install his system anyway. He needed the data to be able to improve his software.

All facial recognition companies wanted real-world photos and video of people taken with surveillance cameras, particularly in situations such as at an airport, where people's identities were being checked, so that they would have a robust database of faces paired with names to improve their software—eventually to the point where it would actually work.

Bruce Schneier, the technologist who had criticized the surveillance of Super Bowl fans, later coined a term for countermeasures such as this, which "provide the feeling of security instead of the reality": security theater. Face scanning would not actually help identify bad guys; its only possible benefit was deception, making travelers feel less anxious and deterring anyone with nefarious intent who believed the facial scanning could be used against them.

Representatives from the ACLU continued to rail against the technology in the media and at security conferences. They expressed frustration that anyone was in favor of the systems. They pointed to the deployments at the Super Bowl and in Ybor City, Tampa, where no one had been successfully identified for arrest. "We have had zero percent increase in the number of bad guys caught," an ACLU lawyer said at a biometric conference in New York City at the end of 2001. "Not a single arrest. That's amazing. What are we doing here? Why should we even debate the privacy issue if there is no security benefit?"

The *St. Petersburg Times,* which had covered the Super Bowl snooping closely, ran a story about the surge of interest in facial recognition after 9/11. Vendors, including Atick and Colatosti, said they were besieged by calls from law enforcement, office buildings, malls, and schools. The story expressed concern that the institutions were seeking out technological snake oil: "While company officials say the technology is as accurate as fingerprints, one government study said the technology is virtually useless in crowd situations." A former army colonel who served as an ACLU board member told the reporter that the organization worried about the harm that flawed technology could cause. "What we fear is that not only will it not identify people in the database, eventually it will be used to harass someone innocent," he said.

Colatosti wasn't saying it publicly but he knew his company's system had at least one glaring problem. It had to abandon a pilot project in South Africa because its technology failed utterly to distinguish among Black faces.

The GAO report ordered by Dick Armey after the "Snooper Bowl" came out in March 2002, into an entirely changed world, one that was worried less about the chilling effects of the technology than about how to make it better. The report revealed that the Department of Defense first began researching and developing the technology in 1987, the year that the paper that inspired Matthew Turk's work came out, and that since then, the federal government had spent a grand total of $47 million on it. The other big spender, besides the Defense Department, was the Department of Justice, which had looked at using it at the southern border in 1994 to "identify suspect criminal aliens." The report's spending numbers predated the terrorist attacks, but the agency noted anecdotally that government interest in the technology had increased significantly.

Later that year, in November, the GAO reported that facial recognition technology was being rapidly adopted by government agencies but estimated that at least $50 million should be spent to assess how well it worked. Seventeen states were using Colatosti's system to protect against identity theft. The State Department was testing facial recognition technology to vet visa applicants. Iceland and Australia had installed facial recognition systems at their international airports, and Atick's system had been piloted at four airports, in Boston, Dallas, Fresno, and Palm Beach, with dismal rates of success.

The largest deployment of facial recognition, though, according to the GAO, was in England, in an "unsafe neighborhood" called Newham Borough in East London. In 1998, Atick's system had been installed on twelve of the area's three-hundred-some CCTV cameras, comparing the faces of passersby to a police database of around a hundred convicted street robbers. It had led to a decrease in street crime, leading to favorable public approval ratings, yet it had not led to any arrests. Its presence alone had evidently scared off would-be criminals. The security theater worked.

Even Tom Colatosti found use cases like Newham's off-putting. He thought that Atick was the one overhyping the technology and that facial recognition should be deployed only in places where there was a security threat, not in the middle of cities scanning the faces of anyone who walked by. He considered that "creepy."

Creepiness aside, the pilots weren't reassuring as to the technical effectiveness of the face-scanning systems. So policy makers in the

United States, who had begun to wonder about the quality of what the government was buying, turned to NIST, a federal lab whose mission was measurement.

NIST, THE NATIONAL Institute of Standards and Technology, was founded in 1901, when American industry was booming, to help standardize the marketplace and ensure that everyone in the country was measuring things such as time, weight, length, light, and power the same way. The agency even conducted sting operations. A black-and-white photo from 1905 shows two Seattle-based officers in suits and bowler hats standing amid a pile of fraudulent measuring devices they had just confiscated.

It was a unique federal agency in that it employed not just policy people but scientists and engineers. They worked on novel measuring devices such as a balloon-borne tool to forecast the weather and, during World War II, a proximity fuse so missiles could explode before they hit the ground, causing more damage than they would have by exploding upon impact.

The agency first took note of facial recognition as a technology worth measurement in the 1990s, in the post-Turk era, when it began popping up more frequently in academic papers. Researchers were making varying claims about their success in getting a computer to recognize a face, but it was apples and oranges. They were all doing their own thing in their own labs with whatever modest little collection of faces they had at hand. NIST needed a way to judge their performances, and the best way to do that was with a huge collection of known faces.

So a scientist named P. Jonathon Phillips, who worked for the army and then for NIST and DARPA, went about gathering them. He funded a professor at Virginia's George Mason University to recruit more than one thousand people to sit for extensive portrait sessions, their faces captured from a variety of angles, to create a face database called FERET. The weaselly acronym incorporated the *F* and the *E* from FacE, *RE* from REcognition, and *T* from Technology. People in the software world call these thousands of photos, each clearly tied to a particular individual, "labeled data"; they are key for both training algorithms and testing how well they work.

FERET became the benchmark against which NIST could measure facial recognition programs. It could do a head-to-head test—literally—to see which company excelled. In 2000, P. Jonathon Phillips ran the first Facial Recognition Vendor Test, giving each company's algorithm the same fairly simple challenge: judging whether two photos, both taken in a studio with ideal lighting, were of the same person.

It was a great idea, but there was a problem: Phillips didn't have the power to force facial recognition vendors to submit their algorithms; all he had was the allure of a contest and the chance for a vendor to prove that its product was the best on the market. NIST emailed invitations to twenty-four facial recognition software vendors, but only five companies took part, and only three of them managed to finish going through the photos in the seventy-two hours allotted. It was a laughably tiny contest but monumental at the same time. There was now a ruler with which to measure this new technology.

Phillips and his colleagues wrote a seventy-one-page report that came out in February 2001, the month after the Snooper Bowl. It was pretty useless for a lay reader, so technical as to be inscrutable, with the authors writing that their report wasn't meant to be a "buyer's guide." The test didn't reveal which system was best, even though that is typically the *point* of a test. Perhaps that was to avoid hurting the feelings of the few companies that had volunteered to participate. There was no best. None of them was particularly good.

If a person in a photo wasn't looking straight at the camera, mug shot style, the systems had a hard time even detecting that there was a face present. None of the systems was ready to be rolled out to surveillance cameras monitoring football fans, London shoppers, or air travelers. But that was happening anyway.

THE STATE DEPARTMENT was one of the first agencies to grant NIST access to real-world portraits for its test, rather than ones taken for experimental purposes, in the form of visa application photos from more than thirty-seven thousand Mexican citizens, with at least three photos of each person.

Ten vendors participated, and this time NIST was explicit about who excelled. The top three companies, including Atick's, correctly matched two photos of the same person 90 percent of the time—but only if the photos had been taken indoors. If the person had been outdoors, where the sun had created unexpected shadows on their face, the algorithms worked, at best, only half the time.

The best-use case for the technology was to make sure that someone wasn't applying for a new government document under a false identity. And that is how the State Department used it, to check for duplicates when someone applied for a passport or a visa, "lessening the possibility of a terrorist or criminal being allowed into the United States or receiving a U.S. passport through fraud," as the agency later put it in a privacy assessment. State departments of motor vehicles deployed it similarly, looking for identity thieves trying to obtain multiple licenses.*

As for using the technology for more sophisticated surveillance, the results were not promising. Challenged to find a particular Mexican traveler's face in a database of thousands, the best companies came up with another photo of the traveler as the top result only about 70 percent of the time. Despite how facial recognition is sometimes portrayed in TV and film, most facial recognition systems don't spit out just one matching result for a given photo, instead returning a sometimes lengthy list of possibilities. The algorithms did reasonably well at putting a photo of the right person somewhere in the top fifty on that list, but again, that was under near-ideal conditions: searching headshots taken for a government database, not faces out in the wild captured by surveillance cameras at oblique angles.

The researchers noted in their report that facial recognition worked better for men than women, and that younger people were harder to recognize than older people.† That intrigued one of the authors: Patrick Grother, who eventually took over the testing from Phillips.

*In 2011, a man in Massachusetts sued his local registry of motor vehicles after it suspended his driver's license simply because the facial recognition system deemed him too similar-looking to another driver. The lawsuit revealed that it was not an uncommon occurrence, raising questions about how often the systems may have wrongly flagged the photos of innocent visa applicants, causing them to be denied entry to the country, likely without explanation.

†Face recognition technology still doesn't work as well on children as on adults.

Grother, an English computer scientist who joined NIST in 1990, ran the test every few years with more vendors participating each time. He saw dramatic gains in accuracy by the top performers, but he continued to see differences in how the algorithms performed on men and women and on people of different ages.

Decades later, Grother got a pained look on his face talking about that finding, sighing and pushing his stylish circular glasses up on his head. "We didn't call it bias," he said. "We should have said, 'This is a demographic problem. This is going to impede fairness.' We should have been more proactive."

Why didn't NIST prominently flag the fact that facial recognition worked better for some people than others? One reason may have been that, although facial recognition clearly fared best on men and adults, racial bias wasn't clear-cut. The bias varied from vendor to vendor but seemed linked, primarily, to whose faces had been used to train a given system, as was apparent during two tests the agency ran in 2009, one for the FBI.

After creating national databases of DNA and fingerprints that law enforcement agencies could tap to identify perpetrators who left bodily evidence at a crime scene, the FBI decided to do the same thing with mug shots. To find out which vendor would be best at doing it, the FBI gave NIST the faces of nearly 2 million arrestees. The data set included their ethnicities, allowing Grother to assess how the algorithms performed on people of different races. He and his fellow researchers found a surprising "race effect": "Blacks are easier to recognize than whites for 5 of the 6 algorithms," they wrote. The majority of the algorithms were biased against *white* people.

The authors didn't reflect at length about that, speculating briefly that it may have resulted from the companies' "optimizing performance for an expected demographic." In other words, the vendors knew that they were going to be tested using mug shots, and given the overincarceration of Black Americans, they may have trained their algorithms accordingly.

Around the same time, Phillips compared algorithms made by Western companies to those made by Eastern companies. "Psychological research indicates that humans recognize faces of their own race more accurately than faces of other races," Phillips and his fellow researchers wrote. They reported the same "other-race effect" with the algorithms. Western algorithms were better at recognizing Caucasian faces and Eastern algorithms were better at recognizing Asian faces. This important research did not seem to have the ripple effects in the face recognition community that it should have, and there was little reckoning with how these disparate performances would play out when the algorithms were deployed in the real world.

To make its millions of mug shots searchable, the FBI chose one of the top performers in NIST's test: L-1 Identity Solutions, which had formed through the merger of the two companies formerly run by the rivals Atick and Colatosti. By 2014, the FBI's billion-dollar "next-generation identification system" had facial recognition up and running. Technical documents warned that if a searched person's mug shot existed in the database, it would appear among the top fifty candidates only 85 percent of the time. Yet when a search was run on the system, it included just twenty candidates by default.

There were success stories; facial recognition helped catch a couple bank robbers and a killer using an assumed name. But there was no real-world audit of how well the technology was faring. No one knew how often it accurately identified criminal suspects, how often it failed, or whether it had led officers to finger the wrong person for a crime. The FBI emphasized that any "candidates"—its careful word for the mug shots turned up in a search—were "investigative leads" only and not "identifications."

Facial recognition technology had been deemed good enough to use based on tests under ideal conditions, but no one knew how well it actually worked out in the wild, where it was increasingly being deployed.

CHAPTER 7

THE SUPERCOMPUTER UNDER THE BED

In 2017, Hoan Ton-That was living in "the tiniest room of all time" on East Eleventh Street in Manhattan, paying $1,100 a month in rent. It was the cheapest place he could find in the East Village. He considered it a steal even though he had to use a bathroom shared by everyone else who lived on his floor. The neighborhood was full of these architectural palimpsests, century-old tenement buildings that had once housed impoverished working-class immigrants, now inhabited by writers, artists, computer engineers, and other "edgy creatives," many of whom, like Ton-That, got their work done in nearby bars, coffee shops, and cafés.

Smartcheckr was now a legit registered business, but it didn't have a real office, just Richard Schwartz's address on the Upper West Side acting as a corporate placeholder. So Ton-That spent a lot of time at a Greek café a couple blocks from his building because it stayed open until midnight and had Wi-Fi. He was pretty much on his own, figuring out the technological backbone for Smartcheckr. Charles Johnson was across the country in California, dealing with a new baby and a disintegrating marriage. Schwartz was in New York, but he was the business guy.

Ton-That, on the hunt for faces and tools to analyze them, tackled that challenge the same way he had learned to code as a teenager: by going online and teaching himself what he could. He went on GitHub, a network for computer programmers to share their work, and typed in "face recognition." He followed machine learning experts on Twitter who would reliably point him in the direction of the most promising techniques. He scoured arXiv.org (pronounced "archive"), a website to which computer scientists and engineers posted academic papers about their work.

Looking back years later, he admitted that it had been an absurd way to begin his project. "It's going to sound like I googled 'Flying car' and then found instructions on it," he said, but his journey had intersected with an important moment in the history of artificial intelligence. The smartest people working in the field, who could command seven-figure salaries from tech giants, were pushing to make their work freely available. They wanted it to be "open source."

This is a little hard for a person who has grown up in a capitalist economy to understand. It's as if the person who came up with Coca-Cola decided to share the recipe with the world. But that's essentially what was happening at the time—and the fault lay, to some extent, with Marvin Minsky, the famed researcher from MIT who had declared neural networks a dead end.

Ever since then, a small group of neural network believers had toiled away in spite of Minsky, convinced that he was wrong and that the biggest breakthroughs in the field would come from programs that could teach themselves through trial and error. Most AI researchers thought the neural network researchers were delusional, but those technologists, who included university professors with nerd-famous names such as Yann LeCun and Geoffrey Hinton, were determined. They kept tinkering with their neural networks, going to conferences and publishing papers about their work, in the hope of recruiting others to their technological cause. And eventually, thanks to faster computers, new techniques, and loads more data, their neural networks started to work.

Once they were up and running, neural networks blew away all the other approaches to AI. They made speech recognition better, image recognition more reliable, and facial recognition more accurate. Neural networks would be employed for all manner of tasks: recommending shows to watch on Netflix, populating playlists on Spotify, providing eyes to the autopilot in Tesla's electric cars, and allowing ChatGPT to converse in a seemingly human way. Anywhere a lot of data exists, a neural network can theoretically crunch it.

All of a sudden, that small cluster of renegade academics became the hottest commodity in technology. Google, Microsoft, Facebook, Baidu: All the biggest technology companies in the world threw money at them. The neural network experts accepted the money, but they had spent years publishing their work in an effort to prove to the world that what they were doing was worthwhile. They liked publishing their work, and they wanted to keep doing it, even though they worked for companies that generally sought to keep their technology proprietary. Not only did those new hires have the leverage to do what they wished, there were upsides for the companies. Other technologists read about their techniques, came up with ways to improve them, and then published papers about *their* improvements—which advanced the field overall. It was a virtuous cycle—one that benefited people such as Hoan Ton-That, who got to capitalize on that technological kumbaya from the sidelines.

Ton-That discovered OpenFace—a "face recognition library" created by a group at Carnegie Mellon University and headed by a graduate student who later landed a job in Facebook's AI lab. "We intend to maintain OpenFace as a library that stays updated with the latest deep neural network architectures and technologies for face recognition," the group wrote in a paper about the collection of code. They posted it on GitHub, and that was where Ton-That found it.

If he read the entire page on GitHub, he would have seen a message from the creators: "Please use responsibly!" Beneath it was this statement: "We do not support the use of this project in applications that violate privacy and security. We are using this to help cognitively impaired users sense and understand the world around them." It was a noble request but completely unenforceable.

Ton-That got the OpenFace code up and running, but it wasn't perfect, so he kept searching, wandering through the academic literature and code repositories, trying out this and that. He was like a person walking through an orchard, sampling the fruit of decades of research, ripe for the picking and gloriously free. "I couldn't have figured it all out from scratch, but these other guys, like Geoff Hinton, they stuck with it and it was like a snowball," he said. "There was a lot of stuff we could mine."

When Ton-That didn't understand something he read in an academic paper, he wasn't afraid to exercise his curiosity. He would go to the professors' websites, find their phone numbers, and call them up to ask questions. Some would actually answer them.

But there were limits to what Ton-That could do, both because he wasn't a machine learning expert and because he didn't have the best equipment for the job. Part of the reason that neural networks had come into their own was the development of new hardware, including powerful computer chips called graphics processing units, or GPUs, that had been developed for video gaming but that turned out to be incredibly useful for training deep-learning neural networks. Ton-That couldn't afford state-of-the-art hardware, but luckily for him, he met someone who could access it for free: Smartcheckr's most important early collaborator, a brilliant mathematician named Terence Z. Liu.

After getting a degree in engineering at Nanjing University in China, Liu moved to Ohio to get a doctorate at the University of Toledo. In the spring of 2017, he was at loose ends, finishing up his PhD in computational physics but unsure how he was going to make a living after he graduated. During a job-hunting trip to New York City, Liu met Ton-That, introduced, he said, thanks to "overlapping tech circles." In a café, Ton-That showed Liu what he was working on. The face recognition capabilities were passable at that point but would need to be much more powerful in order to build a company around them.

Liu wanted a software development job at a big firm and was looking to get more practical experience outside of running physics experiments on university supercomputers. Physicists who want to study stars, the Big Bang, black holes, or, in Liu's case, the creation of new materials rely on supercomputers to simulate the universe and its materials. A supercomputer is what it sounds like: a very powerful machine that performs mathematical calculations at mind-boggling speeds. Around that time, *The New York Times* said of the fastest supercomputer that "a person doing one calculation a second would have to live for more than 6.3 billion years to match what the machine can do in a second."

"I needed to put some projects onto my résumé," Liu said of his decision to work with Ton-That. Ton-That said that Liu helped the company's algorithm immensely because he "stole a lot of computer time on their supercomputer where he graduated from." (Liu later claimed he had simply used a desktop workstation that he had helped build. He downplayed its power, saying that people in his department referred to it as "the supercomputer under my bed," though his PhD advisor, when asked about this, said he had never heard of it.)

Either way, it was a mutually beneficial partnership. The work gave Liu the practical engineering experience he needed to round out his résumé, and his crucial early contributions accelerated the development of Smartcheckr's algorithm.

IN THE SUMMER of 2017, after Liu graduated, he relocated from Ohio to New York and was able to spend more time working with Ton-That in his carousel of "office" spaces: a gelateria, a coffee shop, and a ramen joint. With his mathematical prowess, Liu was much better at all the bits of human tinkering that went into fine-tuning Smartcheckr's neural network. Ton-That was blown away by how much its matching improved. By June, the duo was ready to test their system. They just needed a bunch of faces to practice on.

Again, a tech giant was there to assist. In 2016, researchers at Microsoft had released a public data set with millions of photos of "celebrities," explicitly to help people working on facial recognition technology. Most of the people included were actors, but there were also journalists and activists, some of whom were prominent critics of face recognition. They had no idea that their own faces were being used to improve it.

One night, after Liu and Ton-That got their newly coded facial recognition system to digest those millions of "celeb" photos, they spent hours searching hard-to-match faces and trying to stump the system with doppelgängers—in this case actors who looked similar, such as Natalie Portman and Keira Knightley. But the system kept getting it right. Out of millions of photos, it consistently pointed them to the right person. "This is crazy," Ton-That thought. "This actually works."

Liu decided to release what they'd built as a celeb look-alike tool, publishing it on his personal website. "Out of 10 million faces, which celebrities do YOU look like the most? Say no more. Upload photo," the site prompted. It was a modern version of the 1970 World Exposition exhibit but with more modern celebrities: "You are Anna Kendrick!"

A technologist affiliated with dlib, an open-source library of machine learning algorithms, noticed the site in mid-June and was excited about its reliance on freely available software, tweeting, "Check out this cool app."

Liu's involvement marked a turning point for Smartcheckr, when the pipe dream of an incredibly powerful facial recognition app morphed into a reality. The months he spent working with Ton-That left the algorithm supercharged—as good as, or maybe even better than, anything else on the market. Ton-That insists this was accomplished based on open-source techniques alone, an astounding feat for two newcomers to biometric technology.

Once Smartcheckr had this powerful matching ability, the goal was to pair a person to their own face, not a celeb's. And so Ton-That needed normal people's faces. Luckily the internet provided a fertile hunting ground.

Like any avid hunter, he fondly remembers his first big kill: Venmo, the mobile payments platform. Venmo launched in 2009 as an app that users linked to their bank accounts to easily send and receive money between friends. The company's biggest innovation was turning payments into a social network with a news feed. When you signed up, you created a handle for yourself, gave your first and last name, and uploaded a photo. Or you could use "Facebook Connect" to log in, as many users did, and just port over all your information, including your profile photo, automatically.

When you sent a payment to someone else on the network, you included a little message or some emojis to indicate why, turning everyday expenditures into a social currency, commemorating how you spent your money and whom you spent it with. The payment to a friend was displayed to your network or even the whole world, depending on your privacy settings, along with the message you wrote. (But not the amount you sent—*that* would be a step too far.) Venmo grew quickly, attracting millions of users, and was eventually acquired by PayPal.

Performative payment was a success story, but privacy proponents were alarmed by the app, particularly Venmo's decision to make transactions public by default. If users didn't want the world to see their every Venmo action, they needed to change their privacy settings—and it was unclear how many of them realized that. Busy people don't dive into all the settings on an app or read privacy policies. Most people think that the sheer existence of a privacy policy means a company protects their data, but, in fact, the policy exists to explain, in lengthy legalese, how the company may exploit it. It would be more accurately termed a "data exploitation policy."

Because Venmo was public by design, the company had done little to protect against voyeurs and scrapers, much to Ton-That's delight. Scraping can be a challenging exercise. Most websites are written for human eyes in a flexible language called HyperText Markup Language (HTML) that gives a designer the ability to tinker with font sizes, page colors, alignment, and such. But that formatting is confusing junk language to a data-seeking bot. Scrapers have to "parse" the HTML, extracting the treasure they seek. "It's annoying," Ton-That said. "Websites always change, and if they change the design, then you have to retune your scraper."

But that was not the case with Venmo. On its home page, Venmo had an image of an iPhone, and on its screen was the Venmo newsfeed, showing users who publicly paid each other for things: "Jenny P charged Stephen H, 'Lyft.'" "Raymond A charged Jessica J, 'Power and Internet.'" "Steve M paid Thomas V, 'nom nom nom.'"* This wasn't just a little made-up product demo; it was an actual live feed of transactions in real time, written in a machine-readable format, and it included the users' full names and links to their profile photos.

"You could hit this one URL with a web script and get a hundred transactions," Ton-That said of the scraper he built for the site. "And you could just keep hitting it all day. And each transaction was like 'Here's a photo you can download.'"

*These are real Venmo transactions from around the time Ton-That was scraping the site, as captured by the Internet Archive's Wayback Machine, an invaluable crawler that preserves old versions of web pages.

Ton-That designed his computer program to visit the URL every two seconds. It was like a slot machine, but one where he won every time. The faces spilled out with each pull of the lever. It was "kind of crazy," Ton-That said. "Privacy people" had complained about it, he recalled, but it remained that way for years. And Ton-That took full advantage of it.

In early June 2017, Ton-That sent an email to Charles Johnson and Richard Schwartz with a link to "https://smartcheckr.com/face_search," a parallel version of the tool he had built with Terence Liu, this one for normal people's faces. "Scraped 2.1 million faces from venmo+tinder. Try the search," he wrote. Johnson loved it; he told acquaintances that he could take a photo of them and know whom they'd been sending money to on Venmo.

Ton-That said they needed more photos, particularly of people they were seeking to impress, such as investors. "We need to populate it with everyone they know, and their company. Then it will wow them," he wrote. In other words, they needed to rig the game. Ton-That scraped names from Crunchbase, a website that tracked startups and investors, and then ran them through Google Images, grabbing the top twenty images that came up for each person. That way, potential investors trying out the app were guaranteed to get hits for their own faces as well as those of the people they knew.

Accumulating all those faces came at a cost. Photos take up space, as anyone who has ever had their phone run out of memory knows. Ton-That had to keep all the digital photos somewhere. He estimated that hosting them using Amazon's cloud services would require paying $10,000 or more per month, so, perpetually thrifty, he looked into other options. He wound up rigging his own storage solution, based on an open-source system called Ceph, but that was only a temporary fix. Smartcheckr was going to need more money to keep this operation going.

Ton-That emailed two of the richest people that Johnson had introduced him to: Oculus Rift founder Palmer Luckey and billionaire investor Peter Thiel. "I got a prototype working," he told Luckey, who was intrigued enough to ask for more information about it. That was flattering, but flattery wouldn't pay the bills.

Ton-That had more luck with Thiel, whom he had met in Cleveland the year before. He told Thiel that Smartcheckr could now search a billion faces in less than a second. "It means we can find somebody on the social networks with just a face," he said.

Rich beyond measure, Thiel could invest in promising companies the way a normal person splurges on an afternoon latte. Thiel was on Facebook's board of directors, and so might have found this development troubling with its implications for the privacy of the social network's users. But it must not have occurred to him, or he didn't care, because the next month, July 2017, one of Thiel's lieutenants emailed Ton-That to say that Thiel was interested in investing $200,000. Smartcheckr had its first financial backer.

Accepting the money from Thiel required the partners to set up a corporation. Schwartz turned to Samuel Waxman, a partner at the corporate law firm Paul Hastings who focused on startups and venture capital. When he was seventeen, Waxman worked as an intern on Rudy Giuliani's mayoral campaign, and he and Schwartz had kept in touch ever since. Over lunch at a Greek restaurant, the Smartcheckr duo told Waxman about how they had come to build their facial recognition tool. They did not mention Charles Johnson. By the end of the summer, Waxman had helped them register a new corporation in Delaware, called Smartcheckr Corp, Inc., to accept the money from Thiel. This time, the company only had two shareholders: Ton-That and Schwartz.

Johnson had no idea he'd been cut out of the company, though he knew he wasn't contributing much beyond his connections. He was across the country in California, distracted by his family troubles. And he was an obvious liability, given his toxic reputation, including a Google footprint littered with media articles that accused him of denying the Holocaust and associating with white supremacists.

Though Ton-That and Schwartz had decided to cut Johnson out legally, they kept including him on emails about Smartcheckr's plans. Shortly after the Delaware corporation was formed, Ton-That emailed Luckey, cc'ing Johnson, telling him that they'd "officially closed money from Thiel," and asking Luckey yet again if he'd like to invest or partner with them.

Thanks to Thiel, Smartcheckr now had some real money at its disposal. Unfortunately, Ton-That couldn't bring on his algorithm savant, Terence Liu. Liu's plan had worked; the projects he'd done with Ton-That had proven his mettle, and in the fall of 2017 he took a software engineering job at Bloomberg LP. But hope apparently sprung eternal that he would return, because nearly a year after Liu took that job, the company was still telling investors, in a pitch deck, that he was its chief technology officer, supposedly in charge of day-to-day development operations.

Rather than hiring an actual chief technology officer, Ton-That recruited freelance coders adept at web scraping, telling them only that he needed faces and the links that led to them. The search for face bounty hunters sometimes took him into the darker zones of the internet, places where people wouldn't share their real names. One guy said he had scraped Meetup, AngelList, and Couchsurfing and offered to sell Ton-That the photos. But he wanted to be paid in the cryptocurrency Ethereum.

"I had to swap my Bitcoin to buy it," Ton-That said. "And it was good. I wanted to hire him. He said no, but he introduced me to some of his friends."

Ton-That would eventually hire about a dozen contractors from all over the world to hunt faces on the internet for him. He had no idea who some of those people actually were, only how to pay them. The hunters, in turn, were not told why Ton-That wanted the faces.

The whole process was thrilling for Ton-That. He was café-surfing for office space and recruiting lost boys as temp workers, but for the first time during his decade-long slog in the tech startup world, he had finally created an app with real promise and a strong wind in its sails. He had also become a naturalized American citizen. Like many early-stage startup founders, he was in stealth mode, but his need for quiet didn't stem only from his fear of competitors. There was a state law with stiff financial penalties that could put him out of business before his company was up and running.

CHAPTER 8

THE ONLY GUY WHO SAW IT COMING (2006–2008)

In 2006, James Ferg-Cadima noticed that the cashiers at Jewel-Osco, the Chicago grocery store where he shopped, were wearing big green pins that said, ASK ME ABOUT PAY BY TOUCH! At the entrance of the store was a kiosk advertising the service with a big sign that said, FAST. EASY. SECURE.

One day, as Ferg-Cadima was checking out, the cashier invited him to sign up. "No, thank you," Ferg-Cadima said, simply wanting to get his groceries into his tote bag so he could walk home to his apartment near Wrigley Field.

Ferg-Cadima has floppy brown hair and a smiling, round, approachable face, which was perhaps why the cashier kept pushing. "You never have to pull out your credit card or write a check again," the cashier added, scanning and bagging his items. All Ferg-Cadima had to do was scan his fingerprint; it would be connected to his bank account, and then he could use it to pay every time he came in.

Asking him to hand over his fingerprint just to save a few seconds during check-out struck Ferg-Cadima as "rather bold." "Let me see the sign-up paperwork," he said. He took a brochure from the cashier and took it to work the next week to "dig a little deeper."

Ferg-Cadima, then in his thirties, was the son of Bolivian immigrants. After getting his law degree in Washington, D.C., and working on Latino rights at a racial justice nonprofit, he had moved to Illinois to clerk for a federal judge. At the time of the fateful shopping trip, he had just started a new job at the Chicago branch of the ACLU as a legislative fellow, tasked with coming up with legislation that could strengthen constitutional protections.

He knew this might be challenging, given how polarizing his employer was among some state legislators. "There's a handful who would never meet with the ACLU," he said. The progressive nonprofit was founded in 1920, in the aftermath of the First World War, to defend protesters, socialists, anarchists, and labor organizers who were being arrested, and in the case of immigrants, deported, for challenging the government and powerful businesses. The ACLU is famous for taking unequivocal positions on hot-button topics: It has opposed the death penalty, supported abortion rights, and protected the right to free speech, no matter how controversial. ACLU lawyers fought Japanese internment during World War II, defended the right of Nazis and the Ku Klux Klan to hold public rallies, and fought the detention of alleged terrorists at Guantánamo Bay after 9/11.

Ferg-Cadima was on the hunt for an issue that might inspire bipartisan support, something more "noncontroversial" than the ACLU's usual targets. He had no idea he would find it at a Jewel-Osco. Combing through the Pay By Touch paperwork, he tried to figure out whether there were any relevant laws on the books. Were there any rules governing what could be done with a person's fingerprints or other measurements taken of their body—corporal identifiers collectively called "biometrics"?

He found two bills that had recently been proposed but had not yet passed: the federal Genetic Information Nondiscrimination Act, or GINA, to prohibit employers and insurance companies from using Americans' DNA against them; and the Texas Capture or Use of Biometric Identifier Act, which forbade the commercial collection of a Texan's "retina or iris scan, fingerprint, voiceprint, or record of hand or face geometry." The second one was very relevant but it struck Ferg-Cadima as more symbolic than effective because it could only be enforced by the state's attorney general; individual consumers couldn't do anything to protect themselves if their fingerprint or voiceprint was taken and commercialized without their consent.

Ferg-Cadima did some basic research on biometrics and was troubled by how uniquely identifying they were and what could happen if a company didn't take good care of them. When other forms of identification were exposed to the wrong people—a credit card or a password—a person could get a new one, but they couldn't change their face, voice, or fingers, at least not without painful surgery. "If that were compromised, there was no coming back from that," Ferg-Cadima said. He wondered about Pay By Touch. "So who is this keeper of these biometric prints?"

Pay By Touch had been founded in 2002 by a businessman named John Rogers. Hedge funds and a descendant of the famous Getty family had invested hundreds of millions of dollars in the Bay Area startup. Thousands of stores had reportedly installed the startup's fingerprint scanners that millions of Americans were supposedly using.

Ferg-Cadima was troubled to find that the company's founder had a lengthy court history; in Minnesota, where he'd lived before moving to Silicon Valley, he'd been sued repeatedly over unpaid debts and had pled guilty to disorderly conduct after being accused of domestic violence by a girlfriend.

That would turn out to be just the tip of the iceberg. Ferg-Cadima created a Google alert so that he would get an email whenever an article appeared about the company. At the end of 2007, the alert delivered shocking news: Pay By Touch was in serious financial straits and planned to declare bankruptcy. "Then all my red flags went off," he said. "Illinoisans' biometrics had just become an asset in a bankruptcy proceeding in a different state altogether."

Anyone could theoretically buy those fingerprints and do with them what they wanted. What if someone bought them to create a private fingerprint database for law enforcement? Or what if some nefarious actor got access to them and mimicked the prints of high-value targets, *Mission: Impossible* style, to go on a grocery shopping spree?* Ferg-Cadima knew those scenarios would light a fire under legislators, no matter what side of the aisle they were on. Technology and privacy were a bipartisan space. "All the stars aligned," he said.

* A theft of biometric information wasn't completely far-fetched. When hackers later breached the federal Office of Personnel Management, which performs background checks, they made off with the fingerprints of 5.6 million government employees.

More damaging stories came out about the company and how it had run out of money. Most of them were on Gawker's Valleywag. "Founder and chairman John Rogers has gone missing after a tenure marked by what our tipster called a 'spend big, live big, party big, girls, drugs, meals binge of a global scale,'" Valleywag reported. Pay By Touch had stopped paying its employees. Rogers had hired his mother to run Human Resources. Investors sued, alleging in their complaint that the company had suffered as a result of Rogers's "constant abuse of drugs," saying he had offered cocaine to a board member. *This was the guy* Illinoisans had entrusted their fingerprints to? Ferg-Cadima was aghast.

The company also had more traditional business problems: It was making mere cents each time a customer paid with a fingerprint but spending a lot on the fingerprint-scanning hardware. Pay By Touch was having a hard time signing people up. Some Christians demurred, connecting it to the "the mark of the beast" in Revelation, while others, such as Ferg-Cadima, just didn't see the point of it. One former employee said that those "millions of customers"—the ones the company had bragged about to investors and the press—had been purchased when Pay By Touch acquired a company that scanned fingerprints at check-cashing services, to prevent fraud. Consumer adoption for grocery payment had been a struggle.

In the downfall of the troubled company, Ferg-Cadima saw opportunity. He consulted the ACLU's technology team at the organization's headquarters in New York to brainstorm a basic definition of "biometric identifier," and they went with a spin on what was in the Texas bill: "a retina or iris scan, fingerprint, voiceprint, or scan of hand or face geometry."

The rest of the bill was pretty simple: Unless you were a law enforcement agency, you had to get permission from a person to collect, use, or sell their biometric information, and you had to explain the plan for storing and eventually destroying the information. If a person's biometrics were mishandled or used without their consent, that person could sue for damages of up to $5,000, a stipulation called a "private right of action" that would allow citizens to enforce the law in court;

they didn't have to wait for the government to do it. "We wanted to build in concrete, economic incentives for companies to do their part," Ferg-Cadima said.

Ferg-Cadima had the Biometric Information Privacy Act, or BIPA, drafted by early 2008, and then he worked the halls of the state legislature in Springfield, Illinois, recruiting sponsors and supporters. He would not lead with the fact that he was at the ACLU—he knew this would alienate some lawmakers—so instead he would emphasize that it was an important consumer protection bill. He told them about Pay By Touch—that it had Illinoisans' fingerprints and might sell them off, now that it had gone belly-up. The bankruptcy was a hugely effective sales pitch for the law, and the bill quickly picked up bipartisan support. Beyond those pitches to lawmakers and their staff members, Ferg-Cadima didn't want to make a lot of noise or enlist the press. It would be better if BIPA sailed quietly through the legislative process, escaping the notice of any lobbyists who might slow it down.

The bill was introduced on Valentine's Day 2008, and within five months, it had passed unanimously in both the House and the Senate. Texas's version of the law had also passed and was going into effect in 2009, and Washington passed a similar law (but it excluded data generated from photos). Those states' laws, though, didn't have a private right of action. Consumers couldn't protect their own rights and had to rely on their attorneys general to enforce them. Only Illinois's had teeth in it.

The common refrain is that technology moves faster than the law. This was that rare exception.

The company that had been known as Pay By Touch essentially split in two after the bankruptcy. The check-cashing part of the business, with the majority of its users and their fingerprints, took on a new name and continued to use the fingerprints to prevent thieves from cashing other people's checks. The part of the company that had been working with grocery stores became YOU Technology, which changed its focus from *how* people bought groceries to *what* they bought. More valuable than shoppers' fingerprints were their purchase histories, which could be used to generate targeted ads and coupons and help steer their future decisions. That was a far more successful business model than Pay By Touch, and the company was eventually acquired by the Kroger's grocery chain and then sold for $565 million to a North Carolina data broker called Inmar Intelligence.

Ferg-Cadima stayed on at the ACLU after his fellowship ended, becoming a legislative counsel with the organization for a few years. He moved on to other civil rights groups, and held government roles, and though he never worked on another technology privacy law, the work stayed with him. "To this day, I've yet to use FaceID on my iPhone," he said. "Pay By Touch was a cautionary tale, of the risk when you pass your intimate and personal information on to a corporation."

But consumers without a law protecting their biometrics were left vulnerable, especially as the technology for capturing them improved.

CHAPTER 9

DEATH TO SMARTCHECKR

As the 2018 midterm elections approached, Holly Lynch was thinking about going into politics. A petite woman with closely-shorn platinum blond hair, big brown eyes, and a nose stud, Lynch had lived her entire life—outside of four years at Harvard—on the Upper West Side of Manhattan, and that's where she wanted to run.

Lynch was in her early forties and for the previous two decades had worked in advertising, shaping big campaigns for brands such as Dove and Volvo. Her desire to switch gears professionally grew out of a harrowing experience. Two of them, actually.

A few years before, she had developed a grapefruit-sized tumor in her brain. She had treated it with radiation, it had gone away, and she had gone back to living her life. But then it happened again, another huge tumor in her adrenal gland. She went through chemo—the reason for her barely-there hair—and it was agonizing. She wasn't sure she could survive cancer twice.

Her second bout had happened during the 2016 presidential campaign, and she persevered in the hope of seeing Hillary Clinton become the country's first woman president. But although Lynch managed to beat her malignant nemesis, Clinton did not.

Lynch, a feminist and lifelong Democrat, was distraught by Donald Trump's victory. Rather than just complain about the state of politics, she decided to get directly involved. She told her well-connected parents, also Manhattan natives, that she wanted to run for office, and they recommended a few people from New York political circles to consult. Eventually she was introduced to Richard Schwartz.

That was in the summer of 2017. Smartcheckr had just gotten the $200,000 from Peter Thiel and registered as a Delaware corporation, but it was struggling to land more investors. It should have been easy. It had a futuristic technology that worked; an early prototype could link a person's face to their Venmo spending. Even if Smartcheckr had no revenue, there was a good chance that investors would throw some money at it in exchange for equity and the dream of recouping their investment tenfold or more in the future if the product took off. A few of Schwartz's friends from his New York City circles had invested small amounts, but not enough to give them significant runway. The investors they approached were skittish: How would Smartcheckr monetize face searches? Was it legal?

Desperate for revenue, Schwartz and Ton-That launched a short-lived, ill-fated political consulting business—a decision that seems to have destroyed Smartcheckr.

IN AUGUST 2017, Richard Schwartz met Holly Lynch for lunch at a French bistro called Bar Boulud at Lincoln Center. Schwartz was short, a little overweight, with salt-and-pepper hair, and talked so much that Lynch could barely get a word in. He told her about the New York City projects he had worked on and all the players he knew in the city. "I got Giuliani elected," she recalled him saying.

Lynch told Schwartz that she wanted to run in the district where she had grown up: New York's tiny Tenth Congressional District, which ran from the Upper West Side of Manhattan down into Brooklyn. The seat was held by Jerry Nadler, a fierce liberal who had been in Congress for more than two decades. Lynch thought Nadler had been in the job too long and had lost his sense of urgency for change. She also objected to his gender. Fewer than 20 percent of the members of Congress were women; Lynch felt that men had been running the country for too long. "I had just come out of cancer treatment and wanted to wait until 2020," she recalled. "But Schwartz said I should run immediately, because it was the year of the woman."

After Schwartz told Lynch that she could win and offered to be her policy advisor, she decided to put a half-million dollars of her own money toward the race and pay Schwartz $5,000 a month as a retainer.

Looking back later, she thought Schwartz's advice might have been self-serving, that she was just a money ticket and a guinea pig for his technology. She was unnerved when she saw his invoices arrive, requesting payment to a mysterious entity: "Veritas Strategic Partners Ltd." On a W-9, Schwartz had used his home address as Veritas's business address, just as he had for Smartcheckr.

Like Lynch, Schwartz lived on the Upper West Side, and she ran into him one day walking in Central Park. He told her he was on his way to a meeting at Trump Tower, and insisted that they walk there together. As Lynch remembers it, Schwartz rambled on again about all the important people in New York City that he knew and how lucky she was to have him working on her campaign. Lynch was discomforted but thought this was the nature of politics.

It was puzzling to hire a Giuliani acolyte as a consultant for a Democratic primary, but Schwartz sounded so well connected, and he was registered to vote as a Democrat (not an unusual choice among Republicans in the overwhelmingly liberal city where the Democratic primary was often the only election that mattered). Schwartz said that he was working with a computer genius who was a data wiz and could assist the campaign. Schwartz called him "the Prince," because he was descended from royalty and had a name that was difficult to pronounce, and emailed her a brochure for "Smartcheckr Corp Polling & Social Media Services" to explain what they could do.

"Engage each voter in a unique customer experience that results in a committed supporter who turns out on Election Day," it read. Smartcheckr could offer four services: "1. Enriched voter profiles," "2. Hyper-polling," "3. Micro-targeted voter outreach," and "4. Extreme opposition research."

The first three bullet points were basically variations of the same thing: gathering information about people from social media and reaching out to them, directly and through ads. It was pretty run-of-the-mill data mining, akin to the information gathering that just about every company does these days to deliver those eerily well-targeted ads floating between photos of your friends on Instagram.

The last bullet point, though—"extreme opposition research"—was something else entirely. "Oppo research," a term of art in the campaign world, is the collection of damaging information about a political opponent, ranging from unpaid parking tickets to an extramarital affair. The Smartcheckr brochure implied that it could unearth harmful intelligence that others couldn't: "Through our proprietary social-media search technology and our ability to access unconventional databases, Smartcheckr and its corps of professional researchers can generate opposition research that can fundamentally alter the dynamics of a political race."

That was a discreet description of what was possible with facial recognition technology in the internet age: It could find online photos of people that they had never expected to be connected to them. It might turn up a married politician's secret, pseudonymous dating profile; a long-ago college Halloween party where a white candidate had worn blackface; or revenge porn posted by a candidate's scorned ex. It was impossible to predict what compromising material could be linked to a person's face. Schwartz never explicitly mentioned face recognition to Lynch, though, perhaps because he didn't trust the Democrat to keep it secret.

The "corps of professional researchers" was an exaggeration: The company was mostly still just Ton-That and Schwartz, with a revolving cast of contractors. To figure out which policy positions Lynch should carve out, Schwartz hired a polling company, at a cost of $24,000, to survey her potential constituents. In November, after the pollster surveyed four hundred Democratic voters in Lynch's district, Lynch met Schwartz and the rest of the Smartcheckr team to go over the findings. "The Prince" was Hoan Ton-That, whom she described as quiet and well dressed with long hair. And there was another guy, called Douglass Mackey, a Vermonter who was personable and put her at ease. They met at Lynch's apartment and sat on off-white leather settees in the living room, a cozy space designed to look like a Japanese tea room, with vintage kimono silk screens and photographs of cherry blossoms.

The advice from the $24,000 pollster was to focus on the subway system, a perpetual source of dissatisfaction for New Yorkers. The "path to victory" was to paint Nadler as responsible for crowded and delayed subway trains. To get that message out, the Smartcheckr team proposed collecting voters' emails and phone numbers and then finding them on social media in order to scan everything they had posted publicly: photos, videos, messages. That would allow them, they said, to create targeted messages for voters that would convince them to dislike Nadler and embrace Lynch.

"Is that legal?" Lynch asked. They reassured her that it was.

Smartcheckr wasn't the only company trying to sell the idea of political victory through online data crunching. It was a standard offering from political consultants, but the general public did not become fully aware of just how much their demographics and Facebook "likes" were being crunched until a scandal emerged involving the British political consulting firm Cambridge Analytica. That firm had gotten its hands on the data of millions of Facebook users in an unseemly manner—by hiring an academic to create a Facebook personality quiz, much like the ones Ton-That had devised a decade earlier, that sucked up the information of the person who took it, along with the information of all of their friends. Cambridge Analytica claimed it could crunch all this data to deliver perfectly targeted messages calibrated to sway a voter toward one candidate or away from another. When this data grab and the political manipulation it supposedly enabled became public, Cambridge Analytica imploded and Facebook suffered a reputational hit, even though it had stopped letting users give their friends' data away long before the scandal came out. The whole complicated narrative was boiled down to a simplistic takeaway: Facebook had let its users' data be stolen and helped Trump, on whose campaign Cambridge Analytica claimed to have worked, win the presidency.

By comparison, what Smartcheckr was selling was pretty staid. It just wanted to scrape what New York voters had posted publicly on social media. Users might not have offered their information thinking that a political candidate would use it to try to manipulate them, but this is always a risk when we post information online in a public forum.

After meeting with Schwartz, Ton-That, and Mackey, Lynch convened separately with the friends who were part of her campaign team. She wanted a gut check about whether to move forward with Smartcheckr. Lynch had worked in marketing; she was fine with targeting people with online ads and mailings. But directly calling and emailing voters, using information gathered about them from social media, struck her as potentially invasive.

One of her friends did some digging and then FaceTimed Lynch in the middle of the night to say they needed to meet the next day. He had found the Gawker posts about Ton-That, saying he was a hacker who had created a worm.

"I'm very concerned he was in your apartment and had access to your computer," Lynch recalled her friend saying. "I want you to search your apartment and scrub your computer. He might have put listening devices in there."

That might have been an overreaction, but Lynch took it very seriously. She freaked out and wound down her campaign, deciding to wait until 2020 to run. Her team realized Schwartz's company Veritas Strategic Partners wasn't real, or at least wasn't registered as a business in New York. Lynch blocked Richard Schwartz's number, feeling like an inexperienced candidate who'd been swindled by conservatives.

LYNCH DIDN'T KNOW it then, but Smartcheckr had also pitched its questionable consulting services to a candidate on the far right. A midwestern businessman named Paul Nehlen, who described himself to *The Washington Post* as "pro-White," was running for a congressional seat in Wisconsin that same year. On an alt-right podcast, Nehlen said that Smartcheckr had pitched him social media consulting at a price of $2,500 per month. The brochure, which was posted online by the podcast host who interviewed Nehlen, was very similar to the one that Lynch got from Schwartz but more detailed—it explicitly mentioned facial recognition as a tool to research prospective voters. Nehlen told his podcast interviewer that he thought Smartcheckr was trying to sell him "magic beans" but gave it a try anyway, granting Douglass Mackey access to his Facebook account.

After three months passed and Nehlen had nothing to show for it, he pulled the plug on his involvement. The experience left everyone with bad feelings, and in the fallout, Nehlen decided to reveal a secret he had learned: Mackey had an alter ego. He ran the notorious "Ricky Vaughn" Twitter account, which had tweeted incessantly in support of Donald Trump during the 2016 election with racist memes, antisemitic jokes, and outlandish claims about Hillary Clinton. An MIT Media Lab researcher deemed it the 107th most influential Twitter account for election news, just below that of Senator Elizabeth Warren and above that of NBC News. The account had been anonymous until then, and it came as a shock to Twitter politico obsessives that it was run by Mackey, a graduate of the liberal Middlebury College in Vermont.

The far-right blogosphere obsessed over the outing of Mackey and the mysterious company whose services he had been hawking. A "Right Wing Gossip Squad" blogger googled Smartcheckr and connected it to Richard Schwartz, who described himself on LinkedIn as its co-founder and president. The story got traction on the internet cesspool 4Chan, where one commenter asked why Mackey/Vaughn was working with a "Jewish data mining firm." The Google results for "Smartcheckr" were increasingly populated with links to posts about the alt-right and white nationalism.

All of the bad publicity evidently spooked Schwartz and Ton-That, because they decided to get out of politics, kill off Smartcheckr, and try to make it disappear. Schwartz spiked his LinkedIn page and seemingly hired a reputation management firm, judging from the innocuous nonsense that floated to the top of his Google results. A search of his name combined with Smartcheckr subsequently surfaced an obituary for a different Richard Schwartz who had worked as a dressmaker.

Schwartz and Ton-That seemed to need a fresh corporate identity. In December 2017, they registered a new website address for the company, and in the summer of 2018, they amended the corporate filings in Delaware, making their new identity official. The phoenix to rise from Smartcheckr's ashes was Clearview AI.

The company's fresh start had a domino effect. The name change alerted Johnson to the existence of the corporate entity in Delaware that did not include him. He called Ton-That, irate. "Are you trying to Eduardo Saverin me?" he demanded, comparing himself to a Facebook co-founder elbowed out by Mark Zuckerberg in the social network's early days. Johnson made threats, including a possible lawsuit. The conversation left Ton-That shaken and tearful, according to

Schwartz, who was with him when he got the call. Ton-That told Schwartz he could deal with it, but it spiraled on for months.

Johnson had a lot going on in his personal life at the time. Not yet thirty, he was in the middle of a divorce, splitting his time between Texas, where he had moved, and Los Angeles, where his one-year-old daughter still was. He'd been diagnosed with a degenerative eye disease called Fuchs' dystrophy and thought he was going blind. And his reputation was more toxic than ever. When a Florida lawmaker invited Johnson to attend Trump's State of the Union speech, a few news outlets expressed shock that he would bring a "Holocaust denier." Preoccupied with his life coming apart, Johnson hadn't been substantively involved in the company.

But now aware of how precarious his position was, Johnson recommitted himself to the enterprise and did a flurry of introductions that he thought would benefit the rebranded company. He said he could introduce Ton-That to Jack Dorsey, a co-founder of Twitter and the payment platform Square. "No Dorsey for now," Ton-That told Johnson in an email in June 2018. "We can't have the Valley CEOs who run the platforms find out until we have the data."

Johnson kept pushing to be formally recognized as a co-founder, but Schwartz and Ton-That did not see it that way. He talked a big game but he hadn't put sweat equity into their company; he had just made a few key introductions. They were the ones building the product and pitching it to businesses and investors.

In November 2018, they called their lawyer Sam Waxman, who had helped them register as a corporation in Delaware. At Bobby Van's, a steak house near Waxman's office, they told him the whole story. Waxman had never heard of Johnson before, despite working with the company for a year at that point. Johnson was threatening to hurt Clearview unless he got a piece of it, they told him. Waxman suggested that as part of the formal winding down of the original SmartCheckr LLC in New York, they sign a nondisclosure and non-disparagement agreement to keep Johnson quiet.

But his silence would come at a cost. Johnson wouldn't agree to less than 10 percent of Clearview. The shares were split into three trusts for

Johnson, his ex-wife, and his daughter, an arrangement that would later allow Clearview to issue carefully worded statements claiming that Johnson had no involvement with the company. It was "the best of a bad situation," said Waxman, who pointed out that Johnson's shares would get diluted when more investment money came in.

"Let's finish this off so we can move forward," Ton-That wrote to Johnson, who signed the agreement, which included a secrecy clause and entitled Johnson to a 10 percent sales commission on any customers he brought in. Theoretically, Clearview had a clean slate. There would be no detours. It was time to focus on facial recognition technology alone.

Clearview's database was growing. By the end of 2018, the company had collected a billion faces from the internet, meaning it could identify a significant number of people on the planet, some of whom would have several photos of themselves come up in a search. Clearview now had a more robust product with which to court investors and potential customers. And it had first-mover advantage. Facebook, Google, and other big tech companies had vaster databases of people's photos and names, but they hadn't released a tool like Clearview's. Ton-That believed he could beat them to the punch and reap the rewards. But the reason the Silicon Valley giants hadn't released their own version of a facial recognition tool to identify strangers wasn't that they couldn't build it; it was that they were afraid to.

PART II

TECHNICAL SWEETNESS

CHAPTER 10

THE LINE GOOGLE WOULDN'T CROSS (2009–2011)

In 2009, Google released "Goggles," a revolutionary new spin on the company's famous search engine. The engineers behind Goggles made a cutesy YouTube video to explain what their creation could do.

"Until now, the only option for web search has been typing or speaking. Now you can search by taking a photo," said Hartmut Neven, a quantum physicist who spoke English with a strong German accent. Google had acquired his startup, Neven Vision, after its strong performance in the NIST Face Recognition Vendor Test. Neven appeared in the YouTube video as a cartoon version of himself with a deep cleft in his chin and sunglasses perched on top of his head.

Neven explained that you could point your phone at something, such as the Golden Gate Bridge, a Frida Kahlo painting, or a book called *The Fabric of Reality,* and Goggles would tell you about it in the form of relevant links from around the web.

It sounded cool—revolutionary, even—but when people actually started playing with Goggles, they found it underwhelming. Yes, they could take a photo of a book and find out where to buy it online, but that was a search done just as easily using words. Goggles didn't have the magic of a *true* visual search, with the ability to identify a strange caterpillar in the backyard. The underlying technology wasn't there yet.

The promotional video admitted that, noting that Goggles didn't work well yet on things such as food, cars, plants, or animals, but the animated Neven promised that it would get better. "As this technology advances, we will be able to do more cool things," he said. "Like suggesting a move in a chess game or taking a picture of a leaf to identify a plant."

A year and a half later, in early 2011, a tech journalist from CNN visited Neven to fact-check that prediction. Neven wasn't based at Google's Bay Area headquarters. He worked instead in Southern California, out of the company's small beachside office in Santa Monica. The journalist snapped a photo of Neven squinting at the camera, his lanky body stretched out in a deck chair. Just like his cartoon self, he had sunglasses perched on his head.

Neven told the journalist that Goggles had lived up to its promise and that it could now do something shocking: "Goggling" someone's face would bring up other photos of them available online. The feature was ready to roll out, but Google was worried about the public's reaction and was trying to come up with a way to give people control over whether their faces would be searchable. "People are asking for it all the time," he said. "But as an established company like Google, you have to be way more conservative than a little startup that has nothing to lose."

There seemed to be no laws regarding the use of biometric information in California, but Google's lawyers had concerns. While testing internally, Neven's team had to get explicit consent from colleagues before "goggling" their faces. Still, Neven admitted that some people "are rightfully scared of it." "In particular, women say, 'Oh my God. Imagine this guy takes a picture of me in a bar, and then he knows my address just because somewhere on the Web there is an association of my address with my photo,'" Neven said. "That's a scary thought. So I think there is merit in finding a good route that makes the power of this technology available in a good way."

The concerns of the vulnerable are not always so top of mind in Silicon Valley or even at Google. Years earlier the company sent cars with cameras mounted on their roofs out on public roads to create Street View. It was a cutting-edge feature of Google's digital maps that allowed a user to pick a spot in the world and see a 3D photo of what it would look like if they were standing there, peering around.

Google hadn't considered that some people might be horrified that anyone in the world, with a click of a button, could virtually stand in front of their house. The Borings, a couple in Pennsylvania, sued the company for taking photos of their driveway, alleging invasion of privacy and trespassing, but the slow-moving lawsuit did little to change the company's mind. It was only after pushback from European privacy regulators that Google offered a salve for people who didn't want strangers on the internet prowling virtually outside their homes. Google created a blurring option—initially for people who lived in Germany, where the company had faced the biggest backlash—to pixelate their houses in Street View photos to make them unrecognizable.

Hundreds of thousands of Germans took advantage of this privacy measure—but it backfired when the blurs became beacons. Pro-tech vigilantes sought out the blurred homes in the real world, the locations easily ascertainable from Google Maps, and egged them, leaving notes in their mailboxes that read, "Google's cool." Those who chose privacy over progress thus became the villains. Evidently, there would be no hiding in this new rabidly transparent world.

Now, according to Neven, Google was again working on a product that would radically change the idea of privacy in a public space. Goggles with Face View would mean that anyone on the street could potentially snap a photo of your face and know who you were. That was building on technology that Google had first released in 2008 on its photo storage site, Picasa, in the form of a friend-tagging feature.

"Picasa's facial recognition technology will ask you to identify people in your pictures that you haven't tagged yet," TechCrunch had reported at the time. "Once you do and start uploading more pictures, Picasa starts suggesting tags for people based on the similarity between their face in the picture and the tags you already put in place for them."

The most tedious task in artificial intelligence was gathering heaps of "labeled data," those examples that a computer needs to parse in order to learn. With Picasa, Google had come up with a clever way to outsource that tedium. Picasa users dutifully labeled their friends' faces when they uploaded party pictures, vacation photos, and family portraits. Unpaid and for fun, they helped train Google's facial recognition algorithm so that it could more easily link different photos of the same face. Now Google was considering releasing the fruits of that labor to the world in the form of a real-life face-finding feature.

Though people had grown accustomed to uploading a photo of a friend and getting a suggestion to "tag" them, what Neven was proposing was something entirely new: the power to put a name to a *stranger's* face. In March 2011, the CNN journalist published his article, with a chilling headline: "Google Making App That Would Identify People's Faces."

When other journalists reached out to the company for information about the forthcoming tool, Google's spokespeople said that the story was "the inventions of the reporter" and not something the company was actively pursuing. Neven, 350 miles away from headquarters, had evidently stepped out of line, making remarks that had not been vetted by Google's robust corporate communication team.

A FEW MONTHS later, Walt Mossberg and Kara Swisher interviewed Eric Schmidt, Google's then chairman, at an annual conference they organized called All Things Digital.

Schmidt, with gray, thinning hair, pockmarked skin, and small eyes engulfed by owlish glasses, was calm and measured as Mossberg and Swisher bombarded him with tough questions about the company's privacy practices and the kind of information Google had about its users. When Swisher asked Schmidt about Goggles and facial recognition, he was surprisingly candid. "We built that technology, and we withheld it," he said. "As far as I know, it's the only technology that Google built and, after looking at it, we decided to stop." Mossberg leaned forward, looking shocked, and asked why. He was understandably incredulous that Google had found a line it wouldn't cross.

After consumers got hooked on Gmail, Google's free email service, the company began to mine private messages to generate relevant ads. When Google launched an ill-fated social network called Buzz, it mined Gmail again—to see whom people emailed most frequently—and then connected them publicly, exposing intimate relationships. And beyond putting people's houses on the internet, Google's Street View cars had been secretly sucking up data, including passwords and private emails, from open Wi-Fi networks as they passed businesses and homes. Since its founding in 1998, Google had been on a mission to expand what was knowable by human beings. Why stop at their faces?

Schmidt didn't provide a lot of detail in response to Mossberg's question. He referred vaguely to an earlier comment he had made in the interview about an "evil dictator" who might use it to identify citizens in a crowd. "People could use this stuff in a very, very bad way," he said. "As well as in a very good way," he hastily added.

But even as Schmidt was onstage, telling two pit bull reporters that universal facial recognition was something Google would never, ever release, the company was negotiating to buy Pittsburgh Pattern Recognition. Nicknamed "PittPatt," the company had developed a facial recognition product that had recently been used to identify college students, without their knowledge, by searching for their faces on Facebook.

IN THE LATE 1990S, a young engineer named Henry Schneiderman, who was getting his PhD at Carnegie Mellon, set out to help computers see faces better. His advisor was the renowned computer vision professor Takeo Kanade, who had processed those photos from the Osaka World Exposition.

At the time, most facial recognition researchers used NIST's FERET database. But Schneiderman thought the photos in that database were *too* good: portraits of willing participants, in high resolution, with perfect lighting and a uniform background. "It's simple to find the face," he said, "because the face is the only thing there." That wasn't what computers would encounter in the wild.

To be useful in the real world, computers would need to identify people in candid poses with dim lighting; in grainy surveillance videos; and on streets, looking away from the camera. Schneiderman thought the research community had put the cart before the horse. A computer needed to be able to find a face before it could recognize it.

One of Schneiderman's fellow grad students had done some work in the area already. He had used simple (though complicated at the time) neural network–based techniques to break a digital photo down into tiny component parts and analyze them for patterns that indicated the presence of a face. Given a bunch of photos with faces and a bunch of photos without faces, the neural network was able to work out, fairly well, when a face was present in a photo, but only if it was looking directly at the camera, not when it was tilted or in profile. And Schneiderman's classmate hadn't been able to figure out why. Anyone who has solved an equation in a math class has had a teacher say, "Show your work." The thing about neural nets is that they don't show their work. They don't explain how they solve a problem. They just solve it, or they don't.

"A neural net is what they call a black box. It's kind of like magic. You don't really know what it's doing," said Schneiderman. "When they fail, it's hard to understand what to do next, because you don't understand why it's failing." Schneiderman took a different approach, called Bayesian image processing. It worked, and in 2000, he published his thesis paper about face detection. It was filled with successful examples: a photo of a crowd of people, all their faces bounded by white boxes showing that Schneiderman's program had detected them; a photo of a young Hillary Clinton sitting beside Kofi Annan, both of their faces boxed; the tennis player Pete Sampras, kissing a trophy, his face in profile but detected. Schneiderman's work was impressive enough that he was awarded both his PhD and a faculty gig at Carnegie Mellon.

Feeling that his creation was a big deal and deserved an audience beyond other computer scientists, Schneiderman decided to put it onto the internet, in the form of an online tool. People could visit his website and upload a photo, and the software would put a box around any faces it could detect. This may sound unexceptional now, but it was enough of a big deal back then that the website climbed to the top of an influential social news website "for nerds" called Slashdot, where readers upvoted it and commented on how cool it was. His face-finding website got so much traffic that it crashed the web server twice, knocking the whole robotics department offline. A couple of companies reached out, including Kodak, which was hoping to use the software to eliminate red eyes in photos. A science museum requested help creating an interactive exhibit; museum visitors would sort through a box of disguises to see what it took to hide from the software.

Schneiderman was intrigued by the possible commercial applications, so in 2004, he took a leave of absence from the university and founded PittPatt with a classmate and a business professor. For the sake of curiosity, he looked into how much it would cost to launch the company in California's fabled Silicon Valley; office space for a month there cost what PittPatt would pay annually for the renovated cork factory it had chosen in Pittsburgh.

PittPatt got an inquiry from South Africa about using the tech to count people on buses to make sure drivers were turning over all the fares. An animatronic dinosaur exhibit used PittPatt's face detection to wake up its Jurassic robots when children walked by. But PittPatt kept hearing that people wanted face recognition, so it turned its efforts toward that and found that digital video made it easier to do. "You see a person as their facial expression changes, as they move their head," said Schneiderman. "It's more data. You do better, the more information you have."

An online demo had worked well for attracting attention to the original face detection tool, so Schneiderman made another one using *Star Trek* episodes, sixty hours of which were available on YouTube. The PittPatt team made the episodes interactive so that you could click on Captain Kirk's or Mr. Spock's face, and the software would reveal all the other points in the episode where the character appeared. That was an extraordinary achievement in 2008, and it had obvious security applications. If you had surveillance video of a crime, you could use PittPatt's software to go through endless amounts of tape, pinpointing the moment the perpetrator appeared.*

Big tech companies started calling, expressing an interest in acquiring PittPatt. They told Schneiderman that they wanted to be able to steer the technology's development to make sure it would work on a smartphone. Despite some tempting offers, Schneiderman did not sell; he wanted to stay in control of his product.

But he came to regret the decision when the recession hit in 2008 and banks no longer wanted to loan PittPatt money. Schneiderman and two other employees deferred their salaries for months to keep the company going until a knight in shining, government-funded armor arrived. Law enforcement agencies that had licensed PittPatt's algo-

*It wasn't just useful for police; Schneiderman heard from summer camps that wanted to make it easy for parents to quickly sort through camp photo albums to find all the ones containing their kids' faces.

rithm had deemed it a "star performer." The company's survival was important. The Intelligence Advanced Research Projects Activity, or IARPA, the intelligence community's slush fund for promising technologies, came through with grant money to keep PittPatt going.

IN LATE 2010, while Schneiderman was toiling away in the cork factory, some of his former colleagues at Carnegie Mellon wondered just how accurate commercial facial recognition had become. The team of researchers was led by Alessandro Acquisti, a professor from Italy with dark hair and large brown eyes, who had become obsessed with the economics of privacy.

Acquisti was fascinated by the "privacy paradox," the phenomenon of people claiming to care about their privacy yet not understanding what they needed to do to protect it. Acquisti thought people who used store loyalty cards, for example, didn't grasp the consequences of a store seeing the purchase of a pregnancy test. Acquisti wanted to study that kind of behavior in depth and, in 2004, he saw an ideal place to do it, a new site then called The Facebook.

When the internet first came to Americans' homes via modems and AOL CDs delivered to mailboxes, people invented new identities for themselves, signing on with whimsical screen names or "handles." You could escape your real self, a concept cleverly summed up by a famous cartoon of a canine sitting at a keyboard, telling a furry friend, "On the internet, nobody knows you're a dog."

But Facebook, as it came to be called, had convinced people—first college students and then everyone else—to link their online activity to their real names. They filled out profiles with their birthdays, their hometowns, their schools, their employers. They handed over their preferences—their favorite movies, TV shows, bands, and brands—and served up lists of their friends. And, of course, Facebook had induced them to post photos. *So many* photos. By 2010, Facebook users were uploading 2.5 billion photos per month, mainly of themselves, their friends, and their family members (whom they were dutifully tagging). It was the "most appalling spying machine that has ever been invented," as Wikileaks founder Julian Assange memorably put it.

The online social network was free, but only because Facebook users were themselves the commodity for sale. Marketers loved Facebook because it had the eyeballs and data of millions of people, increasing the odds of targeting the right person with the right ad.

For Acquisti, Facebook was an incredible case study at a massive scale of the economics of privacy. He could observe what people were exposing about themselves and determine for himself what insights could be gleaned about them. He was not the only academic who saw Facebook users as the perfect lab rats. A *New York Times* article at the time about how much academics loved Facebook called it "one of the holy grails of social science," a perfect "petri dish," and an unprecedented opportunity to put millions of people "under the microscope."

Acquisti was able to study what young people considered private, for example, by looking at the profiles of Carnegie Mellon students. More than 90 percent shared their profile photos, but only 40 percent shared their phone numbers. They wanted to be seen but not called. Their phone numbers were sensitive, but their faces, evidently, were not. He doubted that they truly understood what they were giving away.

Acquisti wondered what might be possible if he put PittPatt's algorithm to work on those photos. So he and his collaborators downloaded 260,000 photos posted to Facebook by over 27,000 Carnegie Mellon students. Then, over two cold days in November 2010, Acquisti and his collaborators set up a desk on Carnegie Mellon's campus with a $35 webcam and a few laptops. They asked students walking by whether they would volunteer to have their photos taken and complete a survey about their Facebook usage.

While the students answered the survey questions, Acquisti's team used PittPatt's facial recognition algorithm and the photos the students had posed for to scan the Facebook collection for each of their faces. Each search took about three seconds. As they finished their surveys, the student volunteers would get a surprise: as many as ten photos from Facebook that PittPatt had deemed the closest match to their face. Then the researchers would ask them to identify the ones they were actually in. More than thirty of the ninety-three student guinea pigs found photos of themselves at the end of the survey, including one student who was not a Facebook user but whose photo had been uploaded by someone else.

Acquisti and his colleagues were astonished. Their small experiment proved that you could use off-the-shelf technology and online photos to identify people in the real world. But there were caveats: It worked on only one out of three students, and that was while searching for them within a fairly small population. That was an easier technical feat than searching for someone in a database of 500,000 people or a million or a billion. But still, Acquisti was amazed—and disturbed—that the technology was already that good at unmasking strangers. He knew it would only get better as computers got faster and more powerful and that we were on the road to the "democratization of surveillance." The technology wouldn't be limited to government spy agencies or large corporations; anyone with a little technical know-how would be able to do what Acquisti's team had done.

Acquisti kept the discovery secret for nine months because he was presenting it in the summer at a prestigious security conference in Las Vegas called Black Hat, which required presenters to honor an embargo. In the meantime, Schneiderman finally gave in to a tempting offer from a tech giant. In July 2011, a month before Black Hat, PittPatt posted an announcement on its website: It had been acquired by Google.

By the time Acquisti and his colleagues gave their presentation, it was no longer possible to replicate what they had done. Google had taken PittPatt's technology private. That was a trend in Silicon Valley; tech giants were snapping up promising startups to ensure that they stayed at the forefront of innovative developments. The year before, Apple had bought a Swedish facial recognition company called Polar Rose, and similarly shut down its services to outsiders. The Silicon Valley heavyweights were the de facto gatekeepers for how—and whether—the tech would be used.*

---

* Google declined to answer questions for this book, instead providing a statement: "We were the first major company to decide, years ago, not to make general purpose facial recognition commercially available while we work through the policy and technical issues at stake. We also have a very clear set of AI Principles that prohibit its use or sale for surveillance. We encourage strong guardrails, particularly for facial recognition in public spaces, through regulations and other means."

IT WAS ONLY after he'd gone to Google that Schneiderman heard about the results of Acquisti's experiment. He was surprised at how well his own technology performed, and it made him think about the difficulties to come as facial recognition improved. "When is it fair for people to be recognized? When should we have a right to anonymity?" he wondered. "If we draw that boundary, how do we enforce it?"

He didn't have to worry about that in the short term, though. At Google, Schneiderman and his team were hurriedly put to work to find a way to unlock Android smartphones by simply looking at them. Google wanted a working product by the end of the year to make a splash and to help push Christmas sales of the new Galaxy Nexus.

It was trial by fire. The former PittPatters helped create Face Unlock by their deadline, but it was far from perfect. People quickly figured out that you could use a photo of a phone's owner to unlock the device. Google warned its own employees not to use it. "It didn't take off and Google never pushed it," said a former Google employee who worked on the company's privacy team. "Internally they wouldn't let us use Face Unlock for our corporate phones because it wasn't secure enough. So why are we letting consumers use it?"

It would take years, and better smartphone cameras, to perfect this feature. Apple would eventually beat Google to the punch, with a "True Depth" camera that could differentiate between a flat face and a three-dimensional one using infrared technology. Millions of iPhone users would come to think nothing of using 1:1 facial recognition technology countless times per day.

Judging from patents Google filed around that time, it had other ideas for facial recognition. Just as Arbitron had imagined years earlier, one patent application described using the technology to monitor who was sitting in front of a television and whether they were paying attention to it. Google also imagined using it to track what people in a smart home were doing: Who cooked, who cleaned, who played board games? In the vision of the world painted by those patent applications, each of the camera- and microphone-enabled devices in a savvy, internet-connected home would be closely—and constantly—monitoring its inhabitants.

THOUGH GOOGLE'S ACQUISITION of PittPatt put a slight damper on the Black Hat presentation, Acquisti still had an important message to convey: Anonymity was in jeopardy. It would soon be possible not just to find out strangers' names but to associate them with the ever-growing dossiers being created about them online: the websites they visited, their credit scores, their grocery purchases, their dating histories, their political affiliations. And there would be little you could do to stop it. "It's much easier to change your name and declare reputational bankruptcy than to change your face," he said. Our faces were inherently public and could be used as the key to link our online and offline selves.

This is the challenge of protecting privacy in the modern world. How can you fully comprehend what will become possible as technology improves? Information that you give up freely now, in ways that seem harmless, might come back to haunt you when computers get better at mining it. Acquisti made a prediction: By 2021, it would be trivially easy to identify just about any stranger in America by their face alone.

He was wrong. It happened much sooner than that.

CHAPTER 11

FINDING MR. RIGHT

Venture capitalist David Scalzo came across Clearview AI in the summer of 2018. In his midforties, Scalzo had a tech background but had made a career in finance, working at a series of big-name investment banks on Wall Street, including Bear Stearns before it imploded during the 2008 recession. He now had his own small investment firm, Kirenaga Partners. Unlike the big Silicon Valley players, it didn't have billions of dollars to throw around. It had a smaller stack of chips and had to place them wisely. Scalzo's approach was to find a fresh tech startup with a working product and to write a check for $1 million or less, enough to give the founders a little runway to take off from but not enough to coast.

Scalzo had been chatting with an insurance broker about startups, and she mentioned one doing facial recognition. He asked for an introduction and soon had a coffee date scheduled with Richard Schwartz and Hoan Ton-That at the Algonquin Hotel near Times Square.

They gave Scalzo the pitch: Clearview AI could do instant facial identification of hundreds of millions of faces. The three chatted for an hour and a half, and in the following weeks, about what the company wanted to do and why Scalzo should invest in them. Peter Thiel had given them $200,000 the year before, and now they were trying to cobble together a million dollars more to keep going.

They'd managed to find a few customers, all private businesses. They were advertising three product lines: Clearview AI Search, a background-checking tool; Clearview AI Camera, which could sound an alert when a criminal entered a hotel, bank, or store; and Clearview AI Check-In, a building-screening system that could verify people's identities "while simultaneously assessing them according to risk factors such as criminality, violence and terrorism."

One bank had used their tool to screen attendees for a shareholder meeting. Another wanted to use it to identify high-net-worth customers. A real estate firm wanted text alerts whenever someone on a building's "no-fly list" walked into the lobby. A hotel was interested in developing a seamless check-in experience. Nothing was firm yet, but Clearview was small with low overhead.

Though a tool that could identify almost anyone had obvious security applications, they didn't bring up law enforcement as a target market. They were interested only in private companies; some of their promotional material quoted an unnamed bank security executive as saying, "I have plenty of metal detectors. Clearview is like a mental detector."

They wanted their service to be cheap. A few thousand dollars a month? Sixty thousand per year? They wanted to undercut the competition, but they also wanted people to know that they had done something that their competitors hadn't: They had collected headshots of people from across the web, and their algorithm was a thing of beauty with world-class accuracy. Their facial recognition worked under poorly lit conditions. It could identify a person even if he donned a hat, put on glasses, and sprouted facial hair. It could identify a person in a crowd or in the background of someone else's photo. The person could be looking away from the camera. It could tell sisters apart.

Scalzo didn't have to take Schwartz and Ton-That's word for it. Sitting in the lobby of the hotel, Ton-That got out his iPhone, took Scalzo's photo, and then ran it through the app. It found two photos of Scalzo—his headshots on LinkedIn and on the Kirenaga Partners website. Scalzo fell in love with the technology on the spot.

Other investors had the opposite reaction. Months earlier, Ton-That and Schwartz had pitched another New York investor named John Borthwick, who ran a startup incubator called Betaworks with a $50 million fund. The pair struck Borthwick as "very misfitty," Schwartz in a suit, looking like he was there to sell jackets, and Ton-That, young and visibly nervous. The demo backfired. Borthwick was taken aback, both by the Facebook photos the tool surfaced of him and by all the data on him the company had scraped without his permission. He declined to invest and later deleted his Facebook accounts. "You freaked the hell out of this guy," a mutual friend told Ton-That.

Ton-That dismissed the reaction as "future shock." "It's too much at once, and people can't process it," he said.

Schwartz and Ton-That didn't just demo the future for Scalzo; they gave it to him free of charge. It was part of the investment pitch: "Here's the app. Go use it."

SCALZO WAS ONE of many backers, potential investors, and friends of the company to get free access to the private superpower, with no conditions set for how it could be wielded. Though its existence remained a secret to the public at large, it became a party trick for some of the wealthiest and most powerful people in the world. They used it as a conversation starter on dates, an identification tool for business gatherings, and an advantage at poker tables with strangers. As the science fiction writer William Gibson once observed, "The future is already here—it's just not very evenly distributed."

In January 2019, Schwartz and Ton-That flew across the country to California to make their pitch to the deep-pocketed investors of Silicon Valley. They met with Sequoia Capital, the firm Ton-That considered the Harvard of the venture capital world. Desperate for the validation that would come with a Sequoia investment, he gave the app to firm employees, who would go on to run hundreds of searches.

The firm's managing partner, a billionaire named Doug Leone, was an especially avid fan, but when he failed to invest, Ton-That cut him off. In March, Leone texted Ton-That asking how to reactivate his account. "Term sheet," Ton-That replied. Leone ignored the request for an investment but said he'd pay for the app, explaining that he used it for work, to remember people's names when he traveled to other offices.

A month later, Leone texted again, saying he needed the app for an upcoming conference in Sun Valley, dubbed "Billionaire Summer Camp" by the press. "There are a lot of ceos and only first name tags," Leone wrote. Ton-That gave Leone access again, still hopeful to get into his version of Harvard, but all he got was an email from a friend of Leone's who also wanted to use Clearview. His name was Joe, and his email, saying that he might invest, was full of typos. Ton-That forwarded the message to Schwartz, expressing annoyance that Leone not only wasn't investing himself but was telling other people about the app. Schwartz immediately called Ton-That.

"That's Joe Montana," Schwartz told him.

"Who's that?" Ton-That asked.

When Ton-That eventually met the legendary quarterback, he got a photo with him: Ton-That dressed in white and Montana in black, his iPhone tiny in his immense hand. Montana got access to the app but did not become an investor.

John Catsimatidis, a New York billionaire in his sixties with real estate, energy, and media holdings, but most famous for owning the grocery chain Gristedes, tested out Clearview in one of his East Side stores. His security team wanted to keep out repeat shoplifters. "People were stealing our Häagen-Dazs. It was a big problem," he later said.

Gristedes didn't adopt Clearview in his stores, but Catsimatidis used the app one Tuesday evening while he was having dinner at Cipriani, an upscale Italian restaurant in SoHo. His daughter, Andrea, a striking blonde who favors tight dresses and plunging necklines, walked into the restaurant on the arm of a man her father didn't recognize. Catsimatidis asked a waiter to take a photo of the couple and then uploaded it to the app. Within seconds, it identified the mystery suitor as a venture capitalist from San Francisco. "I wanted to make sure he wasn't a charlatan," Catsimatidis said.

He then texted the man's bio to his daughter. Andrea and her date were surprised but thought it was "hilarious."

Catsimatidis had twenty-eight thousand contacts in his iPhone. He joked with the Clearview team that they needed to make a version of the software that came in the form of glasses, so that he could identify people he was supposed to know and not embarrass himself. He didn't invest in the company, but he told them he would buy those glasses if they ever made them.

Ton-That and Schwartz asked people using the app to keep it quiet because the company was in stealth mode, but not everyone complied. Ashton Kutcher spilled the beans when he appeared on a widely watched YouTube series called *Hot Ones,* in which famous guests are interviewed while eating spicy chicken wings. The host asked Kutcher whether in the future "privacy will be the new celebrity." In other words, would it eventually be harder to be an unknown than a known-by-all? Yes, Kutcher said.

"I have an app in my phone in my pocket right now. It's like a beta app," he said midway through the show. "It's a facial recognition app. I can hold it up to anybody's face here and, like, find exactly who you are, what internet accounts you're on, what they look like. It's terrifying."

Ton-That was annoyed. He had originally pitched the app to Kutcher to aid the work his nonprofit, Thorn, did to prevent the sexual exploitation of children. "You almost blew our cover," Ton-That told him.

"I didn't say the company's name," Kutcher sheepishly replied, according to Ton-That.

Kutcher said he used the app only a handful of times, and that he found it "scary." That wasn't a bad thing; in fact it was "usually a good signal for an early-stage company," he said. After the *Hot Ones* episode, Kutcher set up a due diligence call with Ton-That, but he came away from it unwilling to work with or invest in the company. He was put off both by the company's existing investors and something in Ton-That's past that "didn't sit right" with him, though he couldn't later recall what it had been.

He deleted the app, and then later tried to reload it to help a friend identify a stalker from an online photo. "The app didn't work, so I deleted it again," Kutcher said.

One investor told an acquaintance he hoped that his investment would mean he could have his face and his friends' faces removed from Clearview's database, so they couldn't be found. Another, with Hollywood connections, told Ton-That that Kendall Jenner liked his app. Ton-That didn't know who that was, but he was happy when he looked the model up and saw that she had millions of followers on Twitter and Instagram.

He was less enthusiastic when he heard that the potential investor had shown the app to Jeff Bezos. The ruthless billionaire founder and chairman of Amazon was legendary for kneecapping rivals, and Ton-That knew that Amazon had recently developed a deep learning–based image analysis system called Rekognition that, among other things, searched and analyzed faces. "Whoa, dude, what are you doing?" Ton-That thought. To make things worse, the oversharer never invested in Clearview.

Clearview's biggest problem was its misfit founders. Ton-That had a couple of liabilities: the Gawker articles about his being a hacker who had created a "worm" and online photos of him in a MAGA hat hanging out with the Trump crowd—just the sort of photos a tool like his was designed to unearth. Left-leaning investors were put off. And Schwartz, nearing sixty with the fashion sense of an ex-bureaucrat, wasn't the kind of slick startup entrepreneur that investors felt they could trust. Investors who did due diligence found out that the controversial Charles Johnson had ties to the company and that some of the people that Ton-That had hired to work on the technology and scrape faces were from China, setting off national security concerns. Plus, the app was plagued with "regulatory uncertainty," meaning it would almost certainly face lawsuits in its future.

Investors kept turning them down. "In my mind, this thing is technologically derisked. That is just rare. It's built," Ton-That said later, describing his exasperation. "This thing is freaking awesome. It's so cool. It works. It's magical."

Courting investors was a lot like romantic dating. Ton-That and Schwartz met person after person after person, talking up their talents, handing out their magical tool, each time hoping it would work out. But only rarely was there a spark.

With Scalzo, however, there was serious chemistry. Scalzo experimented with the app, trying it on his kids, his friends, whomever he wanted, to find out how well it worked. It would turn up their photos, or photos of people who looked a lot like them, along with links to where the photos had come from. He was impressed not just by the novelty of a global search engine for faces but by how easy it was to use. He knew it would only get better as the company indexed more of the internet and gathered more photos.

He knew it was going to scare people, but he was convinced that they would come around. "Information is power. If we give information to individuals, then they can use their God-given talents in whatever way they choose," he would later say.

They might use them in nefarious ways, but Scalzo was an optimist. He wanted to get the app into the hands of the "moms of America," and everyone else, so that it would become as common to "Clearview a face" as it was to google someone's name. He believed that facial recognition would help people connect in the real world as easily as they were able to connect online. The possible use cases seemed endless to Scalzo, and the duo was trying to raise only a million dollars for the seed round, at a very reasonable valuation of $10 million—meaning it was a deal that could pay off for Scalzo in a big way if things went well. He decided to invest $200,000, a small enough amount that he didn't think it required much due diligence. The product worked; that was all he really cared about.

Scalzo knew the company was navigating uncharted waters and that there would be legal challenges. Clearview had created biometric information for millions of people without their consent and scraped photos from many, many websites, including LinkedIn and Facebook, two powerful companies that had sued over similar behavior in the past.

LinkedIn was, in fact, at that moment, engaged in a high-profile lawsuit against a "people analytics" firm that had had been scraping its site to create a tool called Keeper that could alert employers when a valuable person was thinking about leaving. LinkedIn had alleged violations of the Computer Fraud and Abuse Act of 1986, an antihacking law that forbade accessing a computer "without authorization," because the data was being taken from LinkedIn's servers without permission.* Though the spirit of the law was to protect information stored on private computers, savvy companies such as LinkedIn used

* The federal law has an amazing backstory. It was inspired by a 1980s movie about a teen hacker, played by Matthew Broderick, who breaks into a government mainframe computer connected to the internet, changes a few settings, and almost starts a nuclear war. The Hollywood actor turned president Ronald Reagan freaked out after watching the movie at Camp David and asked his security advisors and members of Congress to look into the possible threat, setting the stage for the creation of the antihacking law.

the letter of the law as a scarecrow that they could plant in the middle of vast fields of publicly accessible data to deter interlopers. The case had made its way to a federal appeals court in the Ninth Circuit and, however it was decided, would set a precedent for other data scrapers, including Clearview.

Scalzo thought that the company's odds were good and that Ton-That and Schwartz were ready for the inevitable backlash when the wider public discovered what they'd done. He saw their push into a legal gray zone as a business advantage. If Schwartz and Ton-That succeeded in commercializing their universal facial recognition app, they'd have the jump on more risk-averse competitors.

CLEARVIEW COBBLED TOGETHER more than a million dollars from a random collective that included Scalzo, some New York–based lawyers and executives in retail and real estate, Ton-That's old mentor Naval Ravikant, a libertarian investor in California named Joseph Malchow, and Hal Lambert, a Texan famed for starting an exchange-traded fund called MAGA ETF that invested only in companies aligned with Republican beliefs.

Lambert liked to demonstrate Clearview to his friends, who would inevitably ask how they could get it. "You can't," he would tell them. At a dinner one night, he showed the app to the Kentucky Republican senator Rand Paul, a staunch privacy defender who had pushed back against government surveillance bills in Congress. Paul was impressed, and relieved to see an American company building technology to rival that being created in China and Russia, but raised questions about the privacy implications.

A month or so after Scalzo invested, he met up again with Ton-That and Schwartz for a celebratory lunch at Benjamin Steakhouse, a pricey restaurant in Midtown Manhattan, just a few blocks away from the Algonquin. They brought along Charles Johnson. It was around the same time that Johnson had found out that his partners had tried to cut him out of Clearview, but if there were any tension among the three men, Scalzo did not pick up on it. Ton-That introduced the red-headed Johnson as another investor in Clearview, though it was Scalzo who picked up the tab for lunch.

Scalzo found Johnson interesting and "wicked smart"; he had insightful, well-researched theories about what society would look like as technology and science accelerated, as computers got more powerful, and as more data could be mined from people's DNA, brains, and digital trails. Johnson talked about technology as a global battle for dominance and said that Clearview was going to help the United States win the face race so that it could identify spies and those who bore the country and its citizens ill will. He claimed to have government and "agency" connections, hinting at ties to the CIA, which he said had the power to choose the people and companies "they want to win."

Johnson was sharp but also a little scary to Scalzo, who would come to think of him as the character Syndrome from the Pixar movie *The Incredibles.** "A little bit of an evil genius," Scalzo said of Johnson. "And you're never quite sure what is true." He couldn't know how apt the comparison would turn out to be. When Johnson's relationship with Ton-That and Schwartz soured, he, like Syndrome, would wreak havoc by sharing the company's secrets with a journalist—me. But that was years away. At that point, the company was still a well-kept secret.

In the fall of 2019, Scalzo was invited to a fundraiser for a Republican running for a Senate seat in Kansas. Peter Thiel was co-hosting it with the conservative media pundit Ann Coulter at his Park Avenue apartment in Manhattan. Scalzo sidled up to Thiel and started talking about the facial recognition app that they had both invested in. Scalzo took a photo of Thiel and other people at the party and then showed him the results: a collection of their photos from elsewhere on the web.

"I'm invested in that?" Thiel asked, according to Scalzo. "That's cool."

"He played it a little coy," Scalzo said. "He's made a thousand angel investments."

Peter Thiel had provided the seed money that allowed the company to sprout and attract other benefactors, such as Scalzo. Without that initial $200,000 from Thiel, the tool might not exist. Yet, if Scalzo's

*Syndrome, also a redhead, devotes his life to murdering superheroes after Mr. Incredible rejects him as a sidekick.

memory is correct, Thiel was barely aware it existed. It was one plant struggling to thrive in a vast field of startups he'd cultivated.

Scalzo kept a somewhat closer watch on his own crops. He checked in occasionally to see how business was going. And soon it started going quite well. Clearview had finally discovered the right customer: the NYPD.

CHAPTER 12

THE WATCHDOG BARKS (2011–2012)

On a wintry Thursday morning in December 2011, privacy advocates, academics, startup CEOs, journalists, and lawyers from the country's biggest technology companies gathered in a cream-colored conference space a few blocks from the United States Capitol in Washington, D.C.

They had gathered for a workshop sponsored by the Federal Trade Commission called "Face Facts." The name made it sound like an intervention program for deviant teens, and in fact that was the spirit of the event. Everyone was there to talk about how to keep an adolescent technology, on the verge of becoming very powerful, from growing up to become a monster.

The workshop was held in a long, narrow room, with the requisite American flag at the front and curtains drawn over the windows, making it depressing in the way that any space without natural light is. There was a reception table with name badges and an agenda for the day. There were no bagels or pastries due to government austerity measures, but a couple of carafes of coffee were brought in at the last minute by Maneesha Mithal, a government lawyer who had paid for them herself.

The money and power in that room were concentrated among the tech giants. Each had a team of lawyers and policy experts in attendance; they moved together in little clumps like schools of fish. No matter their employer, nearly all of the attendees wore the Washington, D.C., uniform: men in dark suits with inoffensive ties and women in blazers.

Facebook and Google were there, joined by fledgling startups such as SceneTap, which had installed facial detection cameras in more than fifty bars in Chicago that kept a running tally of men and women and their ages. Potential patrons could check the SceneTap smartphone app before heading out to see where they were most likely to score.* (It disappeared before class-action lawyers, who likely would have sued it out of existence, discovered the state's relatively new Biometric Information Privacy Act.)

"Back in my day, you had to do a lap around the bar before committing to the optimal bar stool," joked Julie Brill, a lawyer recently appointed by President Barack Obama as a commissioner of the Federal Trade Commission (FTC).

The FTC was created in 1914 with a mandate to protect American consumers from "deceptive, unfair, and anticompetitive business practices." Historically, that included breaking up monopolies, preventing price-fixing, and punishing companies that made outrageous claims about their products.

But the hundred-year-old agency had completely reimagined its mission for the internet age. The practice of exchanging product for personal data fueled an advertising ecosystem that remained lucrative as long as the companies could keep attracting people to their sites. Marketers were paying handsomely for the opportunity to get the right message in front of the right person at the right time. As the saying went, "If you're not paying for it, you're the product."

Protecting citizens from corporate abuse in the Information Age meant protecting not just their hard-earned money but their increasingly valuable personal data. Many countries around the world had created data protection authorities, which focused full-time on information privacy. But in the United States, where freedom of speech and information had historically trumped the right to privacy, nothing like that had been attempted. Instead, in 2006, the FTC stepped into the vacuum with a new, tiny Division of Privacy and Identity Protection and took the position that companies doing shady things with people's information qualified as "deceptive" and "unfair."

From the start, the FTC was on the back foot. A company could

* Yes, for real.

avoid being accused of deception if it included all possible uses of people's data in dense privacy policies—that no one read.* Another way to avoid FTC scrutiny was not to make any specific promises in a corporate privacy policy. Then the company could claim that technically, it hadn't "deceived" anyone: "We didn't tell our users we planned to sell their personal information to the Devil, but we also never told them we wouldn't give their data to the Dark Lord Satan."

The agency had gone after Facebook, Google, and Twitter for their confusing privacy settings and unexpected uses of people's information, but the FTC could only reach settlements with them, in which the companies had to promise not to do anything bad again. Only if a company later violated that promise could the regulator fine it. And its investigations took too long. By the time a privacy settlement with Myspace was agreed on, no one was using the social network anymore. The privacy division was an underresourced sheriff during a data gold rush, but it was trying its best to anticipate all the new ways companies might try to exploit consumer data. Thus this forum on facial recognition technology.

"Face Facts" kicked off with a clip from the Steven Spielberg film *Minority Report,* in which Tom Cruise's heroic detective character, John Anderton, is on the run from police after being framed for murder. As the fugitive detective walks through a mall in 2054, iris-scanning billboards call out to him with enticements. "John Anderton! You could use a Guinness right about now!" one ad shouts. It was science fiction, a dystopia that was supposed to scare people who watched the film, yet there were people in that room who were trying to make a version of it come true.

During the morning session, a lawyer from Intel named Brian demo'd a product the company had released just a few months before:

* In 2008, a couple of academics decided to study how long it would actually take to read all the privacy policies the average American agrees to in a year. Their estimate? More than two hundred hours. That's twenty-five workdays, or a month of nine-to-five reading. To prove how ridiculous it was to expect consumers to read these agreements, one gaming company added to its online terms of service a claim to "the immortal soul" of anyone who placed an order on its site on April Fool's Day.

facial detection software that advertisers could put into digital billboards. Brian ran the software on his laptop, the computer's camera pointed at his face. As he talked, a screen behind him showed Intel's algorithm going to work on him. "The sensor first looks to find that it's a face. And it looks for eye sockets," Brian said. When the software detected Brian's face, a white square appeared over it. "It then looks to other things. It looks to the ears. So, for example, if two ears are showing, there is an eighty-five percent chance that the person is a male," continued Brian, whose own square had turned blue.

An age predictor generously put Brian into the eighteen-to-thirty-five-year-old "young adult" cohort, even though he was thirty-nine. After determining that Brian was a strapping young man, a "male-targeted ad" started playing, a generic car promotion Intel had created for the demo, showing a sedan's front wheels as words appeared above them: "Style . . . Performance . . . POWER."

When Brian covered one of his eyes with his hand, the square disappeared. The software couldn't see his face unless both eyes were visible. (Intel's algorithm evidently wasn't as good as PittPatt's.)

Like any good tech magician, Brian invited a volunteer to come up to test the machine, asking specifically for a female one. "Just so we can show how that works," he said.

A woman with shoulder-length red hair, which covered her ears, dutifully joined him. She looked directly into the camera, and a white square appeared over her face.

"So there it detects that it's a face," said Brian, seeming relieved.

And then the square turned blue. "It actually shows that it's a male face, so there we go," he awkwardly acknowledged, putting his hand in front of the camera in the hope of resetting the algorithm's thinking. He removed his hand, and the square came back. It was still blue. Instead of the shower gel ad intended for feminine faces, the car ad kept playing over and over: "Style . . . Performance . . . POWER."

"So people have asked about accuracy," Brian said, to laughter from the audience. "For gender, we usually have about a ninety-four percent accuracy. For age range, we have a ninety percent or so accuracy rate." He trailed off and quickly ended the demo. Looking back now, it almost seems staged, as if Intel wanted the regulators to think the technology was so bad they didn't need to worry about it.

Brill brushed off the failed demo, choosing instead to let *Minority Report* set the tone for the day's discussion. "If not now, then soon, we will be able to put a name to a face," she said ominously. "For me, this subject brings to mind one of my favorite songs, a song of the Beatles." She was talking about "I've Just Seen a Face," in which Paul McCartney sings about falling in love with a nameless woman at first sight.

"I can't help but wonder if this song might have turned out differently if facial recognition technology had been around in 1965," said Brill, who may not have realized that the CIA and a man named Woody Bledsoe had tried to make that happen.

"What if, when McCartney saw this face, he had instant access to more concrete information about its owner, like her name, her date of birth or place of work, her email address, online purchasing history, or other personal information?" she said. "Let's assume that, like me, she wouldn't have minded if a young Paul McCartney had invaded her privacy just a little bit," she continued, to laughs.

"But what if she did mind? Or what if instead of one of the Fab Four accessing her information, it was an acquaintance that she wanted to avoid?" she asked. Or an insurer that deemed her too high risk to cover based on a photo of her skydiving? Or a potential employer who judged her character based on a picture "from that party freshman year"?

"Or what if the technology confused her for another person?" Brill finished. "We saw this morning how technology doesn't even get the gender of the person it's looking at right. Brian, if you're back from lunch, so sorry."

The audience laughed.

Brill said that it was important for policy makers to keep the nameless girl from the Beatles song in mind as they thought about what rules needed to be put into place "today and in the near future to protect American consumers and competition alike."

"Competition" seemed tacked on, but it was the ever-present specter in any conversation about potential new privacy laws: that they might tie the hands of technology companies, hamper their innovation, and dim the shining light of the American economy.

Academics who had been invited to speak warned that momentous technological change was afoot and the laws of the land were not ready for it. George Washington University law professor Daniel Solove described the legal system as applied to facial recognition as "rickety" and "incomplete," "with a lot of holes" and "little, if any, legal protections." "As a general matter, is the law ready for facial recognition? Not even close," he said.

Alessandro Acquisti described the 2010 experiment he had done at Carnegie Mellon, using PittPatt's technology to match the faces of students to their Facebook accounts. It would get easier and easier to do, he said, "because so many people are uploading good photos of themselves and using their real names." Automated facial recognition might soon be something *everyone* could do. Privacy isn't just about what people know about you, he said, it's about how that knowledge gives them control over you.

Unsurprisingly, the tech companies' lawyers downplayed the privacy concerns, agreeing that using the tech to deanonymize people at large would indeed be scary. But they weren't doing that! they insisted. They were using the technology in completely innocuous ways, helping people tag friends in photos.

There was just one thing that everyone seemed to believe: Companies should not release software that could identify people by their faces alone without their permission. It was rare for tech lobbyists, government bureaucrats, and privacy advocates to agree on anything. Perhaps the unusual philosophical overlap could be codified. But the FTC couldn't actually draft such a law; it could only nudge Congress to do so.

"We kept screaming from the rooftops that Congress needs to enact legislation," said Maneesha Mithal, the lawyer who bought the coffee at the event and who spent twenty-two years at the agency. After the workshop, the FTC issued a thirty-page report with generic best practices for companies interested in face recognition, suggesting that they be "transparent," give consumers "choice," and "build in privacy at every stage of product development." The report warned that "advances in facial recognition technologies may end the ability of individuals to remain anonymous in public places." People walk down the street, eat at restaurants, fly on planes, check in to hotels, work out at gyms, play on playgrounds, go to strip clubs, and browse stores, secure that theirs is a face in a crowd, unlikely to be recognized by those around them. "A mobile app that could, in real-time, identify anonymous individuals on the street or in a bar could cause serious privacy and physical safety concerns," the report said.

This was the privacy watchdog barking, alerting Congress that danger loomed. Would Congress act before it was too late?

CHAPTER 13

GOING VIRAL

Hoan Ton-That and Richard Schwartz had imagined selling the Clearview AI app to private companies: clerks at high-end hotels greeting guests by name, security guards kicking known shoplifters out of retail stores before they managed to squirrel anything away, and doormen at luxury condos easily deflecting unwanted guests. Store visitors would no longer be strangers. A face linked to a person with lots of followers on social media might get preferential treatment. One linked to someone who had once swiped a lipstick might never be able to browse a Sephora again.

But in mid-October 2018, Ton-That and Schwartz met Greg Besson, a security director at a real estate firm. He had once been a cop, and he completely changed the company's trajectory.

"You know what you should be doing?" he said. "You should talk to my old team. They do financial crimes."

Years later, I would ask Ton-That which police department that was. "Somewhere around the greater New York area," he deflected. I pushed him, asking whether it was the New York City Police Department (NYPD), the country's biggest local police force, whose 36,000 officers are tasked with fighting crime in a city of 8.5 million people.

"Yeah, it's not," Ton-That said.

It was.

GREG BESSON HAD been a lieutenant in the NYPD's Financial Crimes Task Force. That meant going after people adept at identity theft, forgery, credit card fraud, and generally getting their hands on money they weren't entitled to.

He retired in 2018 to take a job in the private sector, working for Rudin Management, a real estate firm that owned apartment and office buildings in Manhattan. Rudin's chief operating officer, John J. Gilbert III, was considering investing $50,000 in Clearview, and the real estate firm wanted to do a trial run with Clearview-enabled cameras that could send an alert when an unwanted person, such as a known robber, walked into the building. Schwartz and Ton-That met with Besson as part of Rudin Management's due diligence, and it was then that Besson, impressed by the app's performance, suggested they talk to his old colleagues at the NYPD.

Schwartz, whose brief stint as a state contractor led to allegations of corruption, was initially reluctant to work with the government again. But he overcame whatever aversion he may have had and asked Besson for that introduction. Besson added Chris Flanagan, the Financial Crimes Task Force's commanding officer, to their email thread, using his NYPD address and thus moving the private correspondence into the public record domain.

Schwartz told Flanagan, in his usual manner, about his time working with Mayor Rudy Giuliani and his resulting connections—in this case, former police commissioner William Bratton and former deputy police commissioner Jack Maple, men that he said he had introduced to Giuliani. Known as the "crime-fighting kings," Bratton and Maple had instituted "broken windows" policing, a controversial theory that considered crime to be driven primarily by societal disorder rather than deeper structural problems such as inequality, poverty, and racial oppression. In the 1990s, the NYPD had cracked down severely on low-level offenses—subway turnstile jumping, public urination, graffiti—as a means of discouraging more serious crimes. They even started enforcing a prohibition on "illegal dancing" at bars that weren't licensed as cabarets, a law that had been sitting on the books since the Prohibition era.

Maple was one of the first police officers to tap into the power of big-data analysis. He started analogue, hanging big maps in his office and placing pins at the locations of past subway robberies. He started to see patterns, lots of pins at criminal hotspots. He called them "charts of the future" and said that officers should patrol those zones, rather than wandering aimlessly. It was a classic "past as prologue" playbook, and when it showed promise, Bratton and Maple applied it to the tracking of other crimes. The maps moved from Maple's wall to NYPD computers and eventually to other police departments around the country.

Though it was heralded as a huge success, a system like this can end up entrenching biases about where crime *should* be, such as in low-income neighborhoods or certain racial districts. Hotspots reflect not just where crime has occurred but where officers tend to look for criminal activity. Regardless, Bratton and Maple were legends within the police department and good names to drop. Ton-That and Schwartz set up a meeting for the following Tuesday at an NYPD precinct in Brooklyn.

On the day of the meeting, Schwartz and Ton-That were prepared. They had done a number of these intro-and-demo shows by now, for investors and private industry. And this had once been Schwartz's expertise—unlocking government money for private industry.

The NYPD had an internal face recognition team, but it could only run a perp's photo against a mug shot database, and that wasn't adequate when the perpetrator was from another country or a junior member of a criminal operation without a record. The financial crimes team had many clear photos of criminal suspects from bank counters and ATMs, but no way to figure out who they were. Clearview could help with that.

About an hour into the meeting, Ton-That demonstrated the app for the officers using real-world surveillance photos. A sergeant sent Ton-That six photos taken in the lobby of a bank. Four were ATM transaction photos, showing a man in front of the machine, and two were photos of the man taken by a surveillance camera near the lobby's ceiling.

The demo must have gone well because the financial crimes team decided to do a test run with the tool. It was a repeat of what had happened two decades earlier at the Snooper Bowl in Tampa, Florida. The police were taking a promising facial recognition tool for a free spin to kick its tires and test out its engine. But this time, the tech actually worked.

The app started generating leads right away, not on every single photo they uploaded, but on a significant number of them. An officer wasn't supposed to immediately arrest a person offered up via Clearview, but rather to regard a suggested match as an "investigative lead." An officer could then look for other evidence that suggested that the person had committed the crime, such as checking their bank records for a deposit of the same amount of money that had gone missing.

Pleased, Flanagan ran it up the chain of command at the NYPD to see about signing a contract with Clearview. In early December 2018, Flanagan told Schwartz and Ton-That that senior leadership was impressed with the app's capabilities and that a couple of other departments wanted to try the tool as well. But it needed a few tweaks for crime fighting: Was there a way for the NYPD to keep Clearview AI from getting access to the photos police officers were running through the app? That was sensitive information about criminal suspects and victims; sometimes officers would run mug shots that might later be sealed by court order. The NYPD did not want Clearview to be able to see who they were looking up. But Clearview had not yet built that kind of user privacy into its app.

"We would also need each search logged with our own case numbers and who conducted the searches to avoid misuse on our end," Flanagan wrote. That, Clearview could more easily do. Police officials needed to make sure that officers were on task. There were obvious concerns. What if officers started inappropriately searching for people who hadn't done anything criminal—someone who said something nasty to them, a protester against police brutality, or a pretty woman at a bar? There's actually a term for this last category: LOVEINT, when intelligence officers use their access to surveillance tools to spy on the human objects of their affection. It's happened at all levels of the government, from NSA employees who accessed email and phone data of romantic partners to local police officers who have snooped in DMV databases to get women's home addresses.

Ton-That got to work building the features Flanagan requested, eventually developing an interface to tag every search with a reason and a case number. It was at that point that Clearview signed a vendor policy and a nondisclosure agreement, which was why Ton-That would later deny having worked with the NYPD. The policy had a section on "unpaid demonstrations/pilot projects" that warned that the city saw it as a gift and there was no guarantee that Clearview would ultimately get a contract. Where Clearview had to provide a mailing address, it listed that WeWork office in midtown Manhattan near the *New York Times* building; it briefly rented a space there when Ton-That had decided he needed a more regular office space than East Village cafés.

A week after getting one of the high-ranking detectives set up, Ton-That sent him an email asking how the "tire-kicking process" was going. The detective, who worked at the NYPD's Real Time Crime Center, where large police departments house their most cutting-edge technology, said it was going very well. "We made a few matches on previously unknown individuals," he said.

Emails flew between Clearview and the NYPD as more officers asked to use the app. Ton-That kept making new accounts and then following up with his new users as if they were beta testers. Which, in a way, they were; they were the first officers to use the "Google for faces" to hunt down criminal suspects. And it was working. One officer told Ton-That that Clearview had helped the NYPD catch a pedophile—a tax attorney who thought he was sending his photo to a twelve-year-old girl. When two on-duty cops were photographed drinking while traveling by train to a colleague's funeral, the internal affairs department asked an officer with access to Clearview to run their faces to identify them.

In February, Greg Besson, the company's initial connection to the NYPD, got back in touch. He had shown the app to a high-ranking officer who worked for the Joint Terrorism Task Force, a collaboration between the NYPD and the FBI. Ton-That sent the officer log-in credentials and a link to download Clearview's app to his iPhone. "This application is amazing," the officer told Ton-That the next day.

"Thank you mate!!!" responded Ton-That. "Let me know who else deserves a copy :-)"

And just like that, Clearview made the leap from a local police department to someone who worked with the FBI.

Schwartz, meanwhile, was trying to schedule meetings with decision-makers, to seal the deal, get a contract with the NYPD signed, and get paid. But he didn't seem to be having much luck. Clearview started to generate automated reports for the financial crimes team with its number of daily searches, up to eighty-seven a day, and Clearview's hit rate, to quantify the app's usefulness. The hit rate was hovering around 50 percent and Ton-That wanted to improve it, which meant adding more people to the database. He tapped his gang of global contractors to continue their scraping efforts. The internet was an immense ocean of faces, and they had fished only a small part of it.

"It was definitely getting better over time, more data in the system to match against," a financial crimes officer said. "On many cases, it was a game changer."

But in the end, it wasn't enough. After six months of "testing," and more than eleven thousand searches, the NYPD decided not to formalize its arrangement with Clearview, because of the "perception of it." For the financial crimes officer, that meant catching fewer identity thieves and check forgers. "But with child exploitation or kidnapping, how do you tell someone that we have a good picture of this guy and we have a system that could identify them, but due to potential bad publicity, we're not going to use it to find this guy," he said. "When there's an active threat to human life, not using something like that is, to me, morally problematic."

The NYPD trial paid off in other ways. In early February 2019, an NYPD lieutenant introduced the Clearview duo to Chuck Cohen, a captain with the Indiana State Police, and Ton-That gave employees there trial accounts. Within a month, the Indiana State Police had signed a contract for a one-year, $49,500 subscription to Clearview, making it the company's first official law enforcement customer, with many more to come.

In the following months, Chuck Cohen became an evangelist for the tool, pitching it to other departments and connecting the Clearview team with officials in Florida and Tennessee. Before departing for London, where he was teaching a cyberinvestigation course, Cohen reached out to Schwartz. "I am often asked about emerging criminal investigative capabilities. And, I think that Clearview AI would likely be of interest to some conference participants," he wrote, asking for permission to spread the word.

Clearview also learned about CrimeDex, that national listserv for fifteen thousand financial crimes investigators. One of the CrimeDex ads they sent out, titled "How To Solve More Crimes Instantly with Facial Recognition," directed anyone with questions to a Clearview contact named Marko Jukic (the same Marko who would one day call an investigator in Texas about a pesky *New York Times* reporter). Clearview was on a path to nationwide, and even global, adoption.

But it was still a ragtag crew with rightward leanings. A college dropout who had worked with Charles Johnson, Jukic had expressed radical views in a series of pseudonymous online postings on neoreactionary blogs, arguing in favor of racial segregation, authoritarianism, and violence against journalists who criticized the far-right movement. He later told a journalist that the posts were "exercises in theatrical hyperbole and comedic satire." Now he was helping to sell the police a revolutionary app with the potential to become a powerful weapon for societal control, and it was spreading like wildfire.

Ton-That also adopted a pseudonym at one point. When a Seattle police detective inquired about the security aspects of Clearview and who was behind it, a "Han T" responded to him, using the email address han@clearview.ai, and attached that legal memo Paul Clement had written. "Han T" was a moniker Hoan Ton-That had created because he was worried about what this suspicious detective would think of Clearview if he saw his online reputation as a hacker who had created a worm.

Outside of that constant threat of being googled, Ton-That was elated. Ever since he'd moved from Australia to San Francisco at nineteen, he'd been trying to build an app that people loved. He'd made silly games, photo filters, and video-sharing tools, all of them trying to capitalize on the latest tech fad. None of them had been truly novel. They hadn't had a purpose, and neither had he. But now, at thirty years old, he'd finally built an app that mattered. It was going viral—just not with an audience he had ever expected to serve.

IN MAY 2019, Josh Findley, an agent in the investigative arm of the Department of Homeland Security, received a trove of unsettling images, flagged by Yahoo!, that seemed to document the sexual abuse of a young girl. Findley, who was based in Fairfax, Virginia, at the time, had sixteen years of experience working child exploitation cases. It was traumatic to do the work, he said, but necessary.

Most big tech companies that host email, photos, and videos actively work to prevent the spread of child sexual abuse material, or CSAM. Many use a technology invented by Microsoft in 2009 called PhotoDNA, which can flag the presence of known abusive material previously catalogued by investigators, even if it's been slightly altered. In 2018, Google went a step further, developing an algorithm to identify abusive photos that had never been seen by investigators before, potentially finding victims law enforcement didn't know about yet. Like Microsoft, Google made its technology available to other companies to use.

The images of the young girl had been found in an account based overseas. Investigators hadn't seen the girl's face before, meaning she was an unknown victim who might not have been rescued by law enforcement yet. From a few clues in the photo—such as the electrical outlets in the wall—they guessed that she was based in the United States somewhere. But where?

In the photos Findley had been sent, a man reclined on a pillow, gazing directly at the camera. The man appeared to be white, with brown hair and a goatee. Findley took some screen grabs of the man's face and sent them to an email list of child-crime investigators around the country, hoping that someone might recognize him. An agent with access to Clearview saw the request, ran the guy's face, and sent Findley an odd hit: an Instagram photo of a heavily muscled Asian man and a female fitness model posing on a red carpet at a bodybuilding expo in Las Vegas. The suspect was neither Asian nor a woman. "That isn't our guy," Findley told her.

"Look in the background," she said.

Upon closer inspection, Findley saw a white man at the edge of the photo's frame, standing behind the counter of a booth for a workout supplements company. That was the match. On Instagram, his face would have been half as big as your fingernail. Findley was astounded. He had never used a facial recognition tool capable of doing something like that.

The supplements company was based in Florida. Findley looked on the company's website to see if the guy was an employee, but he didn't see his picture anywhere. He called the company, keeping his questions a bit vague, just in case the guy was the one who picked up the phone. Findley explained that he was looking for someone who had worked at a booth at a bodybuilding expo in Las Vegas and sent the guy's photo along. The person on the phone said he didn't know who it was; it would have been a temporary worker they had brought on, a local.

Soon after he hung up, someone with a different Florida number called Findley with an anonymous tip: That man was named Andres Rafael Viola. Findley found Viola's Facebook account, which was public but hadn't come up in a Clearview search. A room in some of the photos looked familiar. Findley kept searching and found the victim in Viola's network. Viola, an Argentine citizen living in Las Vegas, was arrested in June 2019, within a month of Findley's receiving the abuse photos, and later sentenced to thirty-five years in prison.

"My supervisor said, 'Contact that company, we want a license,'" Findley said. Since then, the Department of Homeland Security has spent nearly $2 million on Clearview.

CLEARVIEW'S SUCCESS WITH law enforcement showed its investors that it had a real market, and it raised another $7 million. That fall, at an international conference for child crimes investigators held in The Hague in the Netherlands, Clearview AI was the hot topic. Investigators called it "the biggest breakthrough in the last decade," and many international agencies asked for free trials to solve crimes against children, including Interpol, the Australian Federal Police, and the Royal Canadian Mounted Police. It was a powerful tool not just for identifying perpetrators but for finding their victims, because it was the first facial recognition database with millions of children's faces.

The police whisper networks were vast. Officers around the world started asking for access. Among the heaviest users during free trials were the National Crime Agency in the United Kingdom, the Queensland Police Service in Australia, the Federal Police of Brazil, the Swedish Police Authority, and the Toronto Police Service in Canada, as well as Ontario regional police services.

Clearview created a marketing document called "Success Stories" that listed cases it had helped solve alongside startling photos from the investigations. It claimed the tool had been used to identify a New York man wanted after a rice cooker bomb scare, a mail thief in Atlanta, and a woman wanted for forgery in Miami. One graphic crime scene photo showed a deceased Black man who had been shot in the head, blood sprayed on the sidewalk next to him, with a red bar over his eyes to create a semblance of propriety. Next to that were the photos of him alive that Clearview had unearthed to demonstrate that the app could identify corpses.

The presentation ended with a page titled "Rapid International Expansion." On a gray map of the world, the United States and twenty-two other countries were shaded Clearview's branded shade of blue. Some raised eyebrows. After making the leap from local police departments to federal agencies, Clearview was now moving into countries with questionable human rights records and well-documented problems with policing, including Colombia, the United Arab Emirates, and Qatar.

Clearview had momentum, finally, but it could still be derailed. The ACLU was sniffing around facial recognition technology again, thanks to another, better known company that had started selling it to the police.

Sometime in late 2017, a privacy activist at the San Francisco chapter of the ACLU named Matt Cagle had taken note of an Amazon press release touting a new addition to the company's computer vision software Rekognition: "You can now identify people of interest against a collection of millions of faces in near real-time, enabling use cases such as timely and accurate crime prevention." The press release mentioned that Rekognition was already being used by law enforcement agencies.

Amazon was only selling an algorithm to do searches; its customers had to supply the faces. What worried Cagle was that it could be integrated into surveillance cameras so that a person in whom the police took an interest could be tracked in real time. When he poked around, he discovered that the city of Orlando, Florida, had just done exactly that in a two-day "proof of concept" pilot. There was an outcry when the ACLU publicized this, and Orlando ultimately abandoned the system.

One of Cagle's biggest concerns was someone being misidentified as a criminal. To demonstrate how that might happen, he and a tech-savvy colleague ran an experiment. They provided Rekognition with twenty-five thousand mug shots and then searched for matches to all 535 members of Congress. They used Amazon's default settings, which surfaced any face the system was at least 80 percent confident was a match. (Amazon would later criticize this choice, saying it had recommended that police change this setting to a percentage in the high 90s.)

In the ACLU's test, Rekognition incorrectly suggested that twenty-eight members of Congress, including the privacy hawk Ed Markey and the civil rights leader John Lewis, might be known criminals. The false matches were disproportionately people of color. The experiment, which caught the attention of lawmakers and garnered a flood of media coverage, cost the ACLU just $12.33. The ACLU called on Amazon to get out of the surveillance business.

Frustrated police officers thought the activists had it wrong, and that facial recognition technology would make policing fairer. They believed that eyewitnesses were more likely to identify the wrong person than an algorithm was. But the ACLU was just getting started. Local chapters across the country launched a campaign to ban face surveillance, pressuring local city councils to forbid their police from using the technology. And it got results. In the summer of 2019, three cities enacted such bans: San Francisco and Oakland in California and Somerville, Massachusetts, outside Boston.

For Richard Schwartz and Hoan Ton-That, controversy was a business problem: They didn't want customers to be afraid to buy their product. They needed help making law enforcement leaders understand just how valuable Clearview could be for their work. To do that, they turned to Jessica Medeiros Garrison, an Alabama lawyer who had headed the Republican Attorneys General Association in Washington, D.C.

The NYPD had helped Clearview disseminate its existence to investigators, but they were low on the decision-making totem pole. An attorney general was essentially a state's "top cop," and could spread the word from on high. A few months after Garrison started, she took Ton-That to a private event for Democratic attorneys general at a Rolling Stones concert, in a private box at Gillette Stadium near Boston.

Ton-That ran demos on the attendees, including corporate supporters and campaign donors. It made people uneasy. “It rubbed people the wrong way that he was taking pictures and pulling up all this personal information,” one attendee said.

Ton-That ran a Clearview search on William Tong, the attorney general of Connecticut. Tong said that the demo made him uncomfortable and that he did his best to get away quickly.

The public still had no idea that police had access to this powerful tool. The benefits were undeniable—it could help solve heinous crimes—but the veil of secrecy drawn around it was problematic. Citizens have a right to know and debate the use of new technologies that redefine the power of the state. Monied investors knew about Clearview. The country’s top government lawyers had seen it. Yet for more than a year, as thousands of police officers around the country deployed it, Ton-That and Schwartz kept the superpower hidden from everyone else.

But not for much longer. Their cover was about to be blown.

CHAPTER 14
"YOU KNOW WHAT'S REALLY CREEPY?" (2011–2019)

The comedian turned politician Al Franken was one of the most recognizable lawmakers in Washington, D.C. His time performing on *Saturday Night Live* had made his broad smile and dimpled jowls famous, and he had put his face on the covers of the bestselling books he had written to skewer political conservatives, including *Lies and the Lying Liars Who Tell Them: A Fair and Balanced Look at the Right.*

After narrowly winning a Senate seat in Minnesota in 2008, he zeroed in on privacy as his signature issue. Franken was ready to hold shadowy tech companies to account and to rein in what they were allowed to do with people's information. To help him figure out exactly how to do that, he turned to a twenty-seven-year-old lawyer named Alvaro Bedoya.

Bedoya had emigrated from Peru when he was five, after his anthropologist father got a Ford Foundation grant to complete his studies in New York. A Harvard grad like Franken, with a law degree from Yale, Bedoya went to work for the senator, hoping to focus on immigrant rights. But when the lawyers on Franken's staff divvied up the big policy issues, Bedoya wound up instead with the privacy and technology portfolio—which he knew nothing about.

"I literally got stuck with it. No one wanted it," Bedoya said.

To bone up on the topic of privacy, Bedoya printed out all the articles from *The Wall Street Journal*'s award-winning "What They Know" series—a two-year investigation into how individuals' seemingly private purchases, movements, and internet activity were being mined, shared, and sold by technology companies and online marketers. Sitting on a Metro train car or waiting for someone at a bar, Bedoya would pull out his folder of articles and read about the way data brokers surreptitiously tracked what people did on the internet and how myriad companies got access to people's location information through their phones. What alarmed him most was how much big tech companies like Google and Facebook knew about their users, and how easily law enforcement agencies could get access to what they knew.

In May 2011, as a result of Bedoya's research, Franken summoned executives from Apple and Google into the Senate to answer questions about iPhones and Android devices creating detailed historical records of people's movements, how long that location data was kept, and who else got access to it. The hearing was popular, so full that they had to turn attendees away. Al Franken sat front and center at the rounded dais with Alvaro Bedoya as consigliere behind him, diligently taking notes. Bedoya had small, rectangular glasses, and a narrow face with the worried expression of a party planner midevent.

"When I was growing up, when people talked about protecting their privacy, they talked about protecting it from the government," Franken said. He expressed concern about how powerful and all-knowing the private sector had become, with no recourse for citizens to protect themselves or even understand what was being done with their data. "The Fourth Amendment doesn't apply to corporations and the Freedom of Information Act doesn't apply to Silicon Valley," Franken bemoaned.

The hearing generated lots of press coverage. "Maybe smart phones are a little too smart," an ABC News reporter quipped. Franken wanted to dig deeper on the attention-getting topic, and asked Bedoya for a new angle for another hearing. None of Bedoya's ideas passed muster, but then, at the end of one meeting with his policy experts, Franken proposed an idea of his own. Just as his team of lawyers was about to leave the room, Franken suddenly stopped and turned toward Bedoya.

"Hey, Alvaro, you know what's really creepy?" Franken said. "Face recognition. That stuff is really creepy."

"Yeah, that is creepy, Al. Let me find out about it," Bedoya replied.

Franken's face was recognized everywhere. He knew what it was like to lose one's anonymity, to walk down the street and be known by everyone who passes, to have anything you do in public come back to haunt you. "He would tell a funny story about how the day that smartphones started to have cameras was the day he lost thirty minutes of every day he was walking around in the street because everyone would stop him for selfies," Bedoya said.

Franken's laser focus on facial recognition was thrilling, and Bedoya dug in eagerly. He read about the FBI's plans for a national face recognition database, Google using facial recognition to unlock Android phones, and Microsoft offering facial recognition as a way to log in to the Xbox. But what really grabbed his attention was what Facebook was doing.

Other corporate giants that had just started experimenting with facial recognition had made the feature "opt-in," meaning consumers had to actively choose to turn it on; few had. In 2010, Facebook, then the most reckless of the tech giants, had chosen a different route. Nothing got people to sign in like an email saying "So-and-so friend tagged 12 photos of you on Facebook," with a tempting little blue box to click that said, SEE PHOTOS. So when Facebook added automatic "tag suggestions," its euphemism for facial recognition technology, to identify friends in uploaded photos, it said it was turning it on for everybody. If you didn't want your name suggested to friends who uploaded your photo, you had to wade into your privacy settings and turn the feature off.

Facebook's imposition of facial recognition on half a billion people had alarmed journalists; one called it "super-creepy" and "terrifying." Facebook thought the uproar was ridiculous; it wasn't unearthing anything unknown—it was just surfacing the names of friends a user already knew. Bedoya disagreed. He thought the company was being far too cavalier with the technology. Facebook, though, had faced little pushback in the United States thus far, as opposed to in Europe, where privacy regulators fought the rollout, saying it was illegal for Facebook to help itself to EU citizens' faceprints without their permission. (Facebook eventually turned off the feature altogether in Europe.)

Bedoya was also concerned about another radical decision Facebook had previously made. It had forced all of its users to make their main profile photo public, a decision that drew criticism from privacy groups, including the ACLU. Facebook had pushed hundreds of millions of people to put at least one photo of themselves on the public internet, and now was normalizing the use of facial recognition technology. It was perfect fodder for a congressional hearing. Franken put one on the calendar for July 2012, but Bedoya ran into a problem: Facebook initially refused to send anyone to testify.

Facebook had just acquired the company that had helped build its facial recognition–based tagging system—Israel-based Face.com—and moved most of the startup's employees from Tel Aviv to Menlo Park. But it wasn't keen to be the sole industry representative at the hearing, responsible for answering all the lawmakers' questions about the private sector's plans for the tech. Bedoya assured the company that there wouldn't be any surprising issues at the hearing—it would be the same questions everyone was already asking—and the company acquiesced. "People think hearings are about getting answers," Bedoya said. "But it's more about the image of Facebook being hauled before Congress to answer hard questions about this creepy technology."

On a Wednesday afternoon, Franken began the hearing with carefully measured opening remarks. "I want to be clear: There is nothing inherently right or wrong with facial recognition technology. Just like any other new and powerful technology, it is a tool that can be used for great good," he said. "But if we do not stop and carefully consider the way we use this technology, it could also be abused in ways that could threaten basic aspects of our privacy and civil liberties."

Among the witnesses asked to testify were Maneesha Mithal of the FTC, privacy economist Alessandro Acquisti, and law enforcement officials, including an FBI information director and a sheriff from Alabama who was rolling out a facial recognition tool to his department. The person from the FBI and the sheriff said that they were only searching the faces of known criminals, relying on mug shots, and not photos collected for driver's licenses or state IDs. "Ordinary, honest people going about their daily business are not of interest to us," the sheriff said. "Our interests are people who are committing crimes, people who are wanted for questioning about crimes."

Though technically true at the time, that testimony from the government officials was misleading. Eventually the FBI would start looking through social media photos and police departments would get access to driver's license photos. They were just *starting* with the faces of people who had gotten on the wrong side of the law.

Facebook's representative at the hearing was a manager of privacy named Rob Sherman. He assured Franken and his fellow lawmakers that Facebook's "tag suggestion" system was identifying only people's friends, not "random strangers." He also said that Facebook would never share its facial recognition system with a third party—say if the police, for example, showed up with a photo of someone they were trying to identify.

Behind the scenes, Bedoya had raised a different concern with a Facebook policy expert. Now that Facebook had made half a billion people put their profile photo publicly on the web, alongside their real name, which Facebook also insisted users provide, couldn't someone come along and download all the photos and create a huge face recognition database to identify strangers? "And I was unequivocally told, with just no shades of gray, 'No, that cannot happen,'" Bedoya recalled later. He was told that the social network had technical protections against scraping in place that limited how much information anyone could download. He later came to regret trusting those claims.

In the hearing, Franken expressed doubt that simply making phototagging easier was Facebook's ultimate goal. "Once you generate a faceprint for somebody," he said, "you can use it down the road in countless ways." He asked Sherman to say "on the record how Facebook will and will not use its faceprints going forward."

Sherman deftly evaded giving a definitive answer. On that Wednesday afternoon in July 2012, he said he did not know what the company would look like "five or ten years down the road." It was difficult, he said, to address a hypothetical.

ONE AFTERNOON, A little less than five years later, in early 2017, in Facebook's headquarters in Menlo Park, California, an engineer named Tommer Leyvand sat in a leather chair with a smartphone strapped to his head. The phone contained a radical new tool known only to a small group of employees.

Leyvand, an Israeli engineer who had joined Facebook in 2016 after a decade at Microsoft, wore jeans, an olive green sweater, and a blue baseball cap. A black Samsung Galaxy smartphone stood on the brim of the cap, held in place by rubber bands. Leyvand's absurd hat-phone was a particularly uncool version of the future. But what it could do was remarkable.

Everyone in the room was laughing and speaking over one another in excitement, until someone asked for quiet.

"It'll say the name," someone said. "Listen."

The room went silent. Leyvand turned toward a man across the table from him. The smartphone's camera lens—round, black, unblinking—hovered above Leyvand's forehead like a third cyclopean eye as it took in the face before it. Two seconds later, a robotic female voice declared, "Zach Howard."

"That's me," confirmed Howard, a mechanical engineer. This was akin to the technology that Google's Eric Schmidt said his company had built but held back. Though the person-identifying hat-phone would be a godsend for someone with vision problems or face blindness, it was also risky, touching on the third-rail privacy issue that had already caused the company reputational and financial pain.

An employee who saw the tech demo'd thought it was "supposed to be a joke." But when the phone started correctly calling out names, he found it creepy. "It felt like something you see in a dystopian movie," he later said.

Facebook founder and chief executive Mark Zuckerberg saw the writing on the wall for social networks. When his users tired of posting vacation pics, hearting Instagram stories, and reading their uncles' political rants, the social network risked becoming Hewlett-Packard, a dinosaur stuck in a previous era of computing. Zuckerberg wanted to be at the forefront of what he saw as the next big technological shift: the embodied internet, a future in which the digital world would surround us rather than just being something we looked at on a screen. He had devoted significant company resources toward research of radical new potential offerings, including a "direct brain interface" to turn thoughts into texts, a device for hearing through one's skin, and smart glasses that would keep the internet always just a glance away.

Zuckerberg became so focused on technology at the intersection of the physical and digital worlds, a space dubbed the "metaverse," that he eventually changed the name of his company to Meta.

With technology like that on Leyvand's head, Facebook could prevent you from ever forgetting a colleague's name, remind you at a cocktail party that an acquaintance had kids to ask about, or help you find someone you knew at a crowded conference.

Facebook had been busy since that congressional hearing Bedoya had pulled together. When Facebook acquired Face.com, one of the employees who moved to Menlo Park was its chief technologist, Yaniv Taigman, a handsome engineer with an easy sense of humor. The algorithm that Taigman had developed in his startup's scrappy office in Tel Aviv could identify one person out of three hundred possibilities with incredible precision, which he thought was as powerful as needed because the average Facebook user had about that many friends at the time. But when he met with Zuckerberg shortly after his arrival, he learned that the company had much larger ambitions.

"We told him we're having these great results," Taigman said. "We asked him, 'Is this good? Is this what you expected?'"

"Let's go to one million," Zuckerberg replied, unimpressed.

Taigman stared at him in disbelief. Facebook's CEO wanted him to scale up automated facial recognition, from identifying one face out of three hundred to one out of one million. "No way," he spluttered. "That's like solving Alzheimer's disease." He explained that the algorithm they currently had was state-of-the-art, that it couldn't be improved, certainly not to that scale. But Zuckerberg didn't care. He ignored Taigman's objections, not even deigning to respond to them, and ended the meeting.

Taigman felt like the miller's daughter ordered by the king to spin hay into gold. It was an impossible challenge. His stress levels skyrocketed. He had entered the meeting almost euphoric and left it bewildered. But then his Rumpelstiltskin appeared. Five miles away from Facebook's office, at Stanford University, an algorithm called SuperVision, created by a Toronto professor and his two graduate students, blew the competition out of the water at an annual computer vision challenge, correctly identifying objects in photos at unprecedented rates. SuperVision used neural networks technology. This was right at that moment neural nets were finally proving their mettle, and Taigman realized he could try using them to recognize faces.*

Given where he worked, getting the training data was easy. His team selected 4,030 Facebook users, whose faces had been tagged in a thousand or so photos, giving them a data set of 4.4 million photos—ample face flash cards for Taigman's algorithmic baby, named DeepFace.† When asked whether there were any rules or privacy limitations regarding whose Facebook photos the team could use, Taigman said only that the photos had been chosen "randomly." The neural network was trained by scoring its analyses of these faces over and over again—"billions of times, trillions of times," said Taigman. When DeepFace got it right, it gave itself the mathematical equivalent of a clap on the back; when it was wrong, the errors led it to recalculate, a process AI researchers call backpropagation.

When DeepFace started to draw even with human facial recognition, able to match two different photos of the same person over 97 percent of the time, Taigman thought there had to be some mistake, that they were accidentally testing with their training set. "I was sure there was a bug," he said. But there wasn't: DeepFace was better than any facial recognition algorithm publicly released, almost better at recognizing faces than humans were.

Taigman and his collaborators stunned the world in early 2014 when they made their work public. Rather than the photos of unsuspecting Facebook users, they used celebrity images to demonstrate how well it worked. For the main example in the paper, Taigman chose Sylvester Stallone, because Rocky had been his hero while he was

* A key component to neural nets technology was powerful hardware; luckily for Taigman, Facebook had a few graphics processing units, or GPUs, sitting around, given to Facebook for free by Nvidia, the main company that made them. It was enough to get him started.

† Why DeepFace? Another term for neural networks is "deep learning," inspired not by a computer's profound thoughts but by the "layers" of a neural network, which allow the network to compute increasingly complex representations of an image. DeepFace was a nine-layer neural network with more than 120 million parameters.

growing up. Given two photos of Stallone, DeepFace could say with 97.35 percent accuracy that it was the same person, even if the actor's head was turned away from the camera and even when the photos spanned decades.* It could tell that Stallone was both the young Rocky of 1976 and the aged one who had come back for *Creed.*

Journalists once again called the company's facial recognition technology "creepy" and "scary good," expressing alarm that it would be applied to Facebook users' photos. Spokespeople at Facebook were quick to reassure journalists that DeepFace wasn't being actively rolled out on the site. "This is theoretical research, and we don't currently use the techniques discussed in the paper on Facebook," a spokesperson told Slate. But less than a year later, Facebook quietly rolled out a deep learning algorithm similar to DeepFace on the site, dramatically improving its ability to find and accurately tag faces in photos. In recognition of what they had achieved and the company's interest in this new development in AI, Taigman and his team were relocated to sit closer to Zuckerberg at Facebook HQ.

"Everything new is scary. My wife is also scared that I will leave her for some robot," Taigman said. "If you're a scientist and innovator, you shouldn't care about it."

It was a pure expression of "technical sweetness," that feeling scientists and engineers get when they solve an intellectual puzzle. Taigman cared most about technical progress; the societal fears in its wake were for others to deal with. When there is a long line of people working on a problem over years and decades, such as with automated facial recognition, everyone in the chain assumes that someone else will reckon with the societal dangers of the technology once it is perfected. In the thrall of technical sweetness, it's difficult to imagine that you are working on something that perhaps should not exist at all.

AFTER GRILLING FACEBOOK and others about facial recognition in 2012, Bedoya began putting ideas together for some kind of federal law to govern the technology. The privacy hearings Franken had held had

*Humans get a comparison like that right 97.53 percent of the time, so we still had a slight edge, but not for long. There were more improvements to come.

created a real sense of momentum. They had decided to get the privacy ball rolling with location information, proposing a law requiring companies to get explicit consent to record someone's movements or to sell them to third parties. The bill passed out of the Judiciary Committee in December 2012, meaning that the full Senate could vote on it. Getting new laws passed was a challenge, but Bedoya felt the wind in their sails.

Then, in 2013, a storm hit. Bedoya was having pizza with friends at Bertucci's on a Thursday night in June 2013 when his BlackBerry started buzzing with news about a secret dragnet surveillance program orchestrated by the National Security Agency (NSA), a U.S. intelligence agency. Journalist Glenn Greenwald had published a blockbuster article in *The Guardian* revealing that a special federal surveillance court had secretly ordered Verizon, the biggest phone carrier in the United States, to give the NSA information about every domestic call that took place on its network. Bedoya went home and spent the night reading everything he could on the subject. Early the next morning, he met with Franken and his legislative director, who looked just as bleary-eyed as Bedoya felt. "Al, we've got to investigate this. It seems huge. We've got to get to the bottom of this," Bedoya said.

Franken replied, speaking very slowly, "There are things . . . that I know . . . that I may not be able to tell you." Because Bedoya was born abroad and most of his family were foreign nationals or naturalized citizens, he had never applied for a security clearance. And now he felt he was looking at his boss across a great divide.

Soon Greenwald's source was revealed to be a twenty-nine-year-old systems administrator named Edward Snowden. Through his work for an NSA contractor, Snowden had gotten access to thousands of the agency's files and given them to journalists, fueling story after story about what the NSA had done to tap into the digital spine of the Information Age, including the collection of email metadata and data siphoned from undersea internet cables.

"The landscape changed," Bedoya said. Franken and his fellow lawmakers turned their attention back to government incursions on privacy rather than the corporate variety. When Bedoya dusted off the location privacy bill more than a year later and tried to get that momentum back, he discovered serious obstacles. No tech company lobbyist had said to Bedoya's face that they hated it, but Bedoya had the sense, even before Snowden came along, that there were powerful forces lining up against him. In March 2014, Bedoya confidentially sent a version of the new and improved location privacy bill to a few people for feedback, including a lawyer who had advocated for privacy legislation at a nonprofit before becoming a corporate lobbyist and advisor to technology companies. Bedoya had been introduced to the lawyer as a person in the tech industry that he could trust.

A few days later, when Bedoya circulated a press release about the bill to a bcc list that included that lawyer, he got a strange response. "This bill is going nowhere," the lawyer wrote. "Earlier this week I re-connected with the group that worked behind the scenes last year to slow/stop this bill. Everyone is still in the same place I believe."

The lawyer quickly followed up, saying he was "totally redfaced" and that his response had been about a different bill entirely. Bedoya sent a polite response, but the email underscored his concern that industry was coalescing to kill the bill.

Even if the email was a genuine mistake, there was other evidence that the tech industry was working against legal protections for location information. A nonprofit based in Washington, D.C., called the State Privacy and Security Coalition—whose corporate members, including Facebook, Google, AOL, Yahoo, Verizon, and AT&T, paid annual "dues" of $30,000 each—existed to ensure that any new state data laws did not prove costly to its members. In 2012, the same year Bedoya and Franken first tried to get a federal location privacy bill passed, the Coalition reported that it "helped prevent potentially damaging location privacy bills from passing this year that would have created significant risk for our members with location-based services and applications."

"They were nice to your face but they were killing your stuff in the background," Bedoya said of the tech companies. The tactic was all the more infuriating because it worked. The once-promising location privacy bill failed to get traction, and Bedoya never wound up putting together a bill to regulate facial recognition technology. Even as the power and accuracy of automated facial recognition improved exponentially, Congress was unable to pass facial recognition rules or any other privacy legislation. Despite being warned repeatedly about what was on the horizon, the federal government failed to act.

The frustrated Bedoya left Franken's staff to create a think tank at Georgetown University focused on the intersection of privacy and civil rights. A few years later, Al Franken himself left the Senate, though not willingly. In 2017, during the global #MeToo reckoning, a series of women accused Franken of touching them inappropriately in years past. Franken denied wrongdoing but resigned his Senate seat under pressure from fellow Democrats, leaving Congress down one lawmaker adamant about the need for privacy protection.

EVEN WITHOUT THE establishment of rules at the federal level, Facebook would later rein in its facial recognition technology. One big reason was the law James Ferg-Cadima had helped to create in 2008, the Biometric Information Privacy Act, which had sat on the books for years until class-action lawyers noticed it. The first seven BIPA lawsuits were filed in Illinois in 2015, with hundreds more to follow.* One of them was against Facebook, which stood accused of illegally taking the faceprints of 1.6 million Illinoisan users when it rolled out its phototagging feature in the state years earlier. Each violation carried a cost of up to $5,000, so the potential damages translated to a stunning $8 billion.

At first Facebook hoped to get out of the lawsuit by arguing that it hadn't harmed its users in any way with its facial templates. Their identities hadn't been stolen. They hadn't lost any money. There was no loss of life or limb, just the imprint of one. But the amusement park Six Flags had made that argument after being sued by a mother whose

* Technology companies became careful not to collect biometrics in Illinois. When Google created an Art Selfie app that let someone take a self-portrait and then have it matched to a famous work of art, people in Illinois discovered that they couldn't use it. The smart security company Nest, also owned by Google's parent company, Alphabet, sold a doorbell camera that could recognize visitors by face and send special alerts, but the feature wasn't offered in Illinois. Sony decided not to sell its $2,900 robot dog, Aibo, in the state because it was designed to recognize people's faces.

teen son's finger had been scanned for entry during a school field trip. The Illinois Supreme Court had shot it down, finding that the law implied that the harm was the violation of a person's biometric privacy. Six Flags gave up and settled its case for $36 million.

Facebook tried another tactic. It hired facial recognition expert Matthew Turk, the MIT grad of *eigenfaces* fame, now a computer vision professor, to suggest that the law, which prohibited the use of a person's "face geometry," should not apply to Facebook. After reviewing the code for Facebook's deep learning algorithm, Turk provided an expert opinion saying that Facebook's system was looking at the individual pixels in an image and creating a mathematical description of the face to do its matching, rather than making a geometric map of a person's facial features. The judge expressed skepticism that that distinction made a difference. And rather than risk a trial in which it could lose billions of dollars, Facebook agreed to pay $650 million in a settlement.

Tommer Leyvand eventually left Facebook to go work on "machine perception" at Apple. The face-recognizing hat he'd helped develop stayed secret. In Facebook's early days, the company's motto had been "Move fast and break things," but after class action lawsuits, government scrutiny, and tangles with privacy regulators around the world, Mark Zuckerberg had officially retired that motto. Facebook was now moving more cautiously. It didn't need to be the first company to come out with a world-changing people-identifying product. It could afford to bide its time until some other company broke that particular taboo.

CHAPTER 15

CAUGHT IN A DRAGNET

The outsider who blew Clearview AI's cover had spent most of his adult life uncovering the myriad ways law enforcement officers spy on us. Freddy Martinez has intense dark eyes, a narrow face, and a nervous speaking style, perhaps as a result of being a professional paranoiac. His wariness of the police, and of how they might abuse their power, stemmed from his own experience at a protest when he was young and working tech support at a high-frequency stock-trading firm in Chicago. Because money had never come easily for Martinez, the son of Mexican immigrants, he appreciated the firm's $50,000 salary, but he found it depressing to assist traders who made money by using powerful software to buy and sell stocks very quickly, acting on information before the rest of the market could process it. It reinforced his feeling that the system was rigged against working-class people like him.

In 2012, a trading firm that had gone billions of dollars into debt in a single day because of a software error was quickly bailed out by its peers. Martinez knew that Wall Street lenders didn't extend the same helping hand to regular people with soul-crushing debt. He himself had $35,000 in student loans, the endless monthly payments forcing him to work a job he hated. "When you have money, you can make money," he said. "When you don't have it, they come after every last cent you owe."

A local offshoot of the grassroots movement Occupy Wall Street had parked its tents three blocks away from his office. The occupiers were pushing for a redistribution of wealth in America, away from the 1 percent of people who control something like a quarter of the country's money. Though Martinez didn't often walk past the Occupy Chicago camp, the movement became a steady drumbeat in his mind. The system was obviously unfair, as if society's levers of power were out of whack.

That spring, NATO held a summit in Chicago to decide what to do about the decade-long U.S.-led war in Afghanistan. The anticorporate protesters of Occupy Chicago also strongly opposed the military-industrial complex, and as they made plans for large-scale demonstrations, alarmed city leaders marshaled a heavy police presence in response.

Martinez decided to join the protest. Dressed in black, he took the train downtown to Grant Park, where people were gathering for a two-mile march to the NATO summit. As Martinez left the train station, he passed a Dunkin' Donuts—just as a large group of police officers emerged, also dressed all in black, but in full riot gear, with great big pads on their fronts and backs. They carried riot shields and extralong wooden batons and wore helmets with thick plastic visors. Martinez thought they looked like *Star Wars* stormtroopers, but what he found most striking was that they were all holding Dunkin' Donuts coffees and dinky pink-and-orange bags presumably full of doughnuts. "It just looked so funny," he said. The police officers were geared up for war, but they'd stopped for a stereotypical doughnut break.

When Martinez lifted his BlackBerry and snapped a photo, two of the officers ran toward him, yelling and swinging their batons. Evidently, they didn't want to be subjected to surveillance themselves. He quick-jogged away, shaken, and continued on to Grant Park, where thousands of people had gathered. After speeches from antiwar veterans, the crowd set off toward the NATO meeting, riot officers lining the route. The march was mostly peaceful until the end, when some protesters got too close to the convention center and officers brought out the long batons; several people wound up in the hospital, and dozens were arrested. Martinez saw the clashes up close, at one point helping to pull a fellow demonstrator away from an officer. He escaped unscathed but galvanized, suddenly aware of the connection between social inequity and the police state; those with wealth and power leaned on law enforcement to crack down on dissent and preserve the status quo because they liked the world just the way it was.

Martinez started befriending more people in the Occupy movement, gravitating toward other activists who were technically minded. One of them had a theory that the police were tracking protesters using a device called a stingray that could trick phones into thinking it was a cell tower; if phones connected to it, cops might be able to log people at the protest and capture data from their devices. The devices were expensive—they could cost over $100,000—and used only by federal agencies, as far as anyone knew, but Martinez decided to look into it. He sent a public records request to the Chicago Police Department, but the department ignored him. He followed up repeatedly but got nothing. So he found a lawyer who was willing to take his case for free and sued. It took six months, but eventually the police were forced to turn over documents that revealed that the Chicago police had been using stingrays for three years.

That city police officers were using those high-tech devices became a national news story, and journalists started interviewing Martinez. Though he was a bit socially awkward—one reporter described him as "fidgeting"—he reveled in the attention he brought to the issue of government surveillance.

MARTINEZ EVENTUALLY MOVED to Washington, D.C., to work for a nonprofit called Open the Government, where it was part of his job to file public records requests. In May 2019, he learned of a hearing on Capitol Hill about facial recognition technology and "its impact on our civil rights and liberties." He watched via a C-SPAN live stream on his computer.

It was a bitterly divisive time in Washington, D.C. The opening speaker was a Democrat who would soon lead the investigation into the Republican president, Donald Trump. He was joined by two Republicans who were the president's biggest allies. But the issue of facial recognition bridged the partisan divide. The technology hits "the sweet spot that brings progressives and conservatives together," even those "on the polar ends" of the political spectrum, said a North Carolina lawmaker who would later become Trump's chief of staff. "We are serious about this and let us get together and work on legislation. The time is now before it gets out of control."

A panel of experts there to brief the congressmen about their concerns included Joy Buolamwini, a Ghanaian-American-Canadian researcher with the MIT Media Lab who had done a high-profile study documenting bias in facial analysis systems. She had discovered the problem while building an "Aspire Mirror" to reflect an inspirational face when someone looked at it, such as that of the tennis star Serena Williams. But the off-the-shelf software she was using had failed to detect her face—until she put on a white mask. When she investigated products from four companies, including Microsoft and IBM, that claimed to be able to predict someone's gender from their face, she found that they worked best for light-skinned men and worst for dark-skinned women. She had concerns about not only the accuracy of facial analysis technologies but people's lack of control over when they might be subjected to them.

"Our faces may well be the final frontier of privacy," Buolamwini said at the hearing. The United States had no restrictions to prevent "private companies from collecting face data without consent," she said in written testimony. She had no idea that Clearview had already done this and at a massive scale.

Also testifying was a lawyer named Clare Garvie who worked for the Georgetown center founded by Alvaro Bedoya. Garvie had been one of his earliest hires, and she had spent her years there studying how law enforcement was using facial recognition. In a 2016 report called "The Perpetual Line-up: Unregulated Police Face Recognition in America," she and her colleagues had revealed that the faces of 117 million American adults were subject to facial recognition searches by law enforcement and that sixteen states included driver's license photos in criminal search databases. Just a week before the hearing, Garvie had dropped a new report about ways in which law enforcement was abusing the technology, including one particularly egregious example: After someone vaguely resembling the actor Woody Harrelson was caught on a surveillance camera stealing beer, the NYPD ran the suspect's face through its facial recognition system and didn't get a hit, so it then ran a photo of Woody Harrelson to see if *that* got a hit.

"My research has uncovered the fact that police submit what can only be described as garbage data into face recognition systems, expecting valuable leads in return," Garvie said at the hearing. "Face recognition is already more pervasive and advanced than most people realize," she warned the members of Congress, "and we should only expect it to become more so."

AFTER WATCHING THE hearing, Martinez got to work right away to find out just how pervasive face surveillance technology was. How many police departments were using it? Who were they buying it from? How much were they paying?

He partnered with Beryl Lipton, a self-described "queen of the FOIA freaks,"* who worked for MuckRock, an organization that helped people request public documents with automated tools that would keep pinging an agency that failed to respond. Martinez and Lipton sent information requests to dozens of police departments, requesting internal communications, marketing materials, training manuals, and contracts or financial documents related to the use of facial recognition. It was a fishing expedition, and Martinez caught a whale.

Late on a Friday afternoon in November 2019, Martinez was sitting in his cramped D.C. office on K Street, about ready to check out for the weekend, when an email arrived from the Atlanta Police Department in response to a request Martinez had submitted four months earlier. Attached to the email was a PDF, and the first words Martinez saw when he opened it stunned him: "Privileged & Confidential, Attorney Work Product."

The memo was just a couple months old, dated August 14, 2019, written by Paul Clement, solicitor general under President George W. Bush, and was titled "Legal Implications of Clearview Technology." It was a letter describing and defending Clearview AI.

* FOIA stands for Freedom of Information Act, the federal law that grants citizens access to public records.

"In the simplest terms, Clearview acts as a search engine of publicly available images," Clement had written, comparing it to Google. Clement said that Clearview's database included "billions of publicly available images," a jaw-dropping number. Martinez knew from the congressional hearing that the FBI had only about 36 million photos in its database. Clearview made America's top law enforcement agency look like a bit player.

The app found people on social media pages and in news article photos, on employment and educational sites, and on other "public sources," according to Clement. Not only was the company selling people's Facebook photos to the cops, more than two hundred law enforcement agencies were already using the app.

Clement had written the memo to reassure potential police customers that they wouldn't be committing a crime by using the app. The tiny, unknown company had hired one of the most famous—and expensive—lawyers in the country to explain why this particular version of face recognition technology was *not* illegal.

But Martinez was from Illinois, one of the few states that, thanks to ACLU lawyer James Ferg-Cadima, required companies to get people's permission to use their biometric information. He was familiar with the law and was certain that what Clearview was doing had violated it.

He couldn't find much about the company online, just a solitary website that carried a vague slogan: "Artificial intelligence for a better world." But the Atlanta Police Department had turned over three other documents about Clearview: a brochure with the tagline "World's best facial-recognition technology combined with the world's largest database of headshots. Stop Searching. Start Solving"; an email with the company's price list; and a purchase order from September 2019 that showed it was shockingly cheap. According to the order, the Atlanta police had agreed to pay Clearview just $6,000 for three yearlong licenses.

Clearview AI had crossed the Rubicon on facial recognition technology and was selling it to law enforcement agencies, and no one else knew it existed.

He looked again at the dates on the documents. They were brand new. Both the legal memo and Atlanta's purchase order for Clearview's tech had been created in the months *after* Martinez had filed his public records request.

Martinez printed out the documents and put them into his backpack. He lived about forty-five minutes from the office and usually took the bus, but that day he walked home, his phone in front of his face, poring over the Clearview PDF. When he got home, he spread the pages out on his bed and read them again. What that little unknown company claimed it could do would mean the end of privacy as we know it.

That was when Martinez reached out to me. I was in Switzerland when I got his email, and it was near midnight. I called him immediately.

CHAPTER 16

READ ALL ABOUT IT

A week after my call with Martinez, I reached out to Jessica Medeiros Garrison, who appeared to be Clearview's chief saleswoman. Her contact information was listed on that brochure given to the Atlanta Police Department.

"I'm a tech reporter at *The New York Times,*" I wrote to her two days before Thanksgiving. I also left her a voicemail message, but she did not respond.

I tried to contact Paul Clement, who had written Clearview's legal memo. I emailed Richard Schwartz, whose name and address were listed in business filings. I reached out to two potential investors—Peter Thiel and Kirenaga Partners. Hoan Ton-That got a voicemail message for the fake Clearview employee he had created on LinkedIn. "I'm looking for someone named John Good," I said.

The small Clearview crew wasn't sure what to do. I later learned that they decided to stonewall, hoping I would lose interest. Only Schwartz responded, with a brush-off. "I've had to significantly scale back my involvement," he wrote. "As soon as I have a chance, I'll see if there's someone who can get back to you."

Ton-That saw via Facebook that we shared a friend, a guy with whom he played soccer named Keith Dumanski. Dumanski did PR for a real estate company and had previously worked in politics. Ton-That hoped that maybe I was working on a basic roundup of companies working in the facial recognition space and not an exposé on Clearview. Dumanski told him that he barely knew me but agreed to reach out to try to gather intel.

Two weeks before Christmas, Dumanski sent me a Facebook message that Ton-That had drafted. When I asked Dumanski for his number, he didn't respond because he had no idea how he'd answer my questions. Clearview realized it needed a professional.

During Schwartz's days with Mayor Rudy Giuliani in City Hall in the 1990s, he had met Lisa Linden, a public relations consultant who specialized in crisis communications—a good person to call when having a PR emergency. When New York governor Eliot Spitzer resigned amid a prostitution scandal, it was Linden whom he hired as his spokeswoman. A Manhattanite to the core, Linden had once discovered a bird's nest in a bush in front of her second home outside the city and found it so novel that she called the local paper to photograph it.

The Monday after Dumanski's failed attempt to elicit information about my aims, Schwartz and Ton-That trekked to Linden's office in Rockefeller Center to ask for advice. They told her that a *Times* reporter had come calling but that they weren't ready to launch yet and asked how to make me go away. When they described my outreach and showed her some of the messages, Linden broke the news to them: There was no stopping the article. But she would help them weather the storm.

IN EARLY JANUARY, after a trip home to Australia for the holidays, Ton-That spent hours with Linden preparing for the questions I was likely to raise. He would be the face of the company while Schwartz stayed in the shadows. On a Friday morning, he and Linden went to a WeWork in Chelsea that they had booked for our meeting. Ton-That wore a blue suit and glasses; Linden brought Tate's cookies, in chocolate chip and oatmeal raisin.

Ton-That told me about himself, and the company, and its extraordinary tool. He declined to name anyone else involved beyond Richard Schwartz, whom he said he had first met at a book event at the Manhattan Institute, a conservative think tank, though he could not remember what the book was when I asked. It came off as strange; and, of course, I would later learn they were actually introduced by Charles

Johnson. I had found business records that showed the company had originally been called Smartcheckr, and that it had been tied in some way to a pro-white political candidate. Ton-That said they had changed the name to Clearview not because of any concerns about Smartcheckr's reputation but because it was better branding. "When you're doing something like face recognition, it's important to have nice colors and not be creepy," he said.

He was proud of what the company had done and said he and Schwartz had made the ethical decision to only scrape public images that anyone could find—3 billion faces at that time. He glossed over the company's narrative, saying it had decided to offer the product only for security purposes because that was the best use case for facial recognition: to catch child predators rather than being used by them. "Every technology has to have a reason to exist," he said. "You can't just put it out there and see what happens."

Clearview claimed in its marketing materials to have one of the most accurate facial recognition algorithms in the world. I asked if it had been audited by a third party or if it planned to hand its algorithm over to NIST for testing. Ton-That said that the company had convened a three-person panel to test the software; it had consisted of a technologist, a former judge, and an urban policy expert. They had done a version of the test the ACLU had conducted with Amazon Rekognition, running the photos of the 535 members of Congress, as well as lawmakers from Texas and California. The panel had determined that Clearview was "100 percent accurate" because it had produced "instant and accurate matches" for all 834 lawmakers. It had been far from a rigorous test, though it worked as a way to tout how much better Clearview was than Amazon Rekognition.

I asked about the alert that had been set up specifically for me and why officers who searched for my face had gotten no results. "It must have been a bug," Ton-That slyly suggested and denied any other knowledge of it. The officer in Texas who had been helping me had gotten a call from "Marko at Clearview" minutes after he searched for my face. That was not a "bug" in the software. But I could not ask Ton-That about the call at that point, because I did not want to get my source into trouble for working with me without official approval from his police department.

"We had a system at the time that gave alerts if people were using it inappropriately," Ton-That would later tell another reporter. A congressional staffer who interrogated the company's chief of security months later about Clearview's ability to see police officers' searches was told that the company's only official monitoring was for accounts with a large number of queries and that the company's CEO, Hoan Ton-That, had personally noticed the searches for my face. Charles Johnson later said he had ribbed Ton-That about blocking my face in the Clearview app, telling him that it came across as "sketchy."

Now Ton-That demonstrated Clearview on me, running my face as well as that of the photographer who accompanied me. Clearview found many photos of both of us. It worked even when I covered the bottom half of my face with my hands. It surfaced a photo of me from almost ten years earlier at a gala in San Francisco, with a person I had been talking to for a story at the time. I hadn't seen the photo before or realized that it was online. It made me think about how much more careful I might need to be in the future about meeting sensitive sources in public places where we could be spotted together and easily identified.

Ton-That pulled up a website called This Person Does Not Exist, which displays AI-generated photos of fake people that look incredibly real. He took a photo of one of the faces and ran it through Clearview. It generated no results.

What Clearview had done was astounding, but not for technological reasons. Companies such as Google and Facebook were capable of building the technology that Clearview had built, but they had regarded it as too taboo to release to the world. The significance of what Clearview had done was not a scientific breakthrough, it was ethical arbitrage. Ton-That and his colleagues had been willing to cross a line that other technology companies feared, for good reason.

You've created a tool that can find all the photos of me online, photos I didn't know about, I said. You've decided to limit it to police, but surely that won't hold. This will open the door to copycats who might offer the tool more widely. Could this lead to universal face recognition, where anyone could be identified by face at any time, no matter how sensitive the situation?

"I have to think about that," he said. "It's a good question."

After the interview, Linden and Ton-That went outside to wait for a car to pick them up. The photographer and I huddled together on the other side of the street, talking about how the interview had gone.

"They're saying it's crazy," Ton-That told Linden as they later recounted the conversation. "And it is crazy. It's the first time in human history that you can identify someone just from a photo."

"After today, your life will be forever changed," Linden warned him.

TON-THAT DECIDED TO subscribe to *The New York Times* so that he could read the story that came out a week after the interview at the WeWork. It was on the front page of the Sunday edition: "The Secretive Company That Might End Privacy as We Know It."

It revealed Clearview's existence, its database of billions of faces, its use by six hundred federal and local law enforcement agencies as well as a handful of private companies, and the efforts it had made to stay hidden from public view. The article also revealed a hint of what might be in the company's future. We got a copy of the app's source code, and my colleague, a technologist named Aaron Krolik, made a fascinating discovery: programming language indicating that Clearview's facial recognition software could be run on augmented reality glasses, specifically products from two companies we'd never heard of before, RealWear and Vuzix. Ton-That downplayed the discovery; he acknowledged designing such a prototype "in our labs" but said that Clearview had no plans to release it.

Hoan Ton-That's photo was featured prominently in the story. He and Clearview were no longer anonymous. It was similar to what had happened to Ton-That a decade earlier, when Gawker had exposed him as the person behind ViddyHo. People recognized him at parties. He got hate mail. But he still felt proud of what he had built and the attention it had generated. "I'm walking clickbait," he said. His desire for fame was greater than his fear of infamy.

Other repercussions were more serious. Privacy hawks in the U.S. Senate—Ron Wyden of Oregon and Ed Markey of Massachusetts—demanded answers. "Widespread use of your technology could facilitate dangerous behavior and could effectively destroy individuals' ability to go about their daily lives anonymously," Senator Markey wrote in his letter to Ton-That.

The attorney general of New Jersey, Gurbir Grewal, barred police in the state from using Clearview AI. That was ironic because Clearview had just put a promotional video up on its site about the role it had played in New Jersey's Operation Open Door, a child predator sting. The video included footage of Grewal at a press conference talking about the arrest of nineteen people. Grewal said he hadn't known that Clearview had been used in the investigation and that he was "troubled" that the company was "sharing information about ongoing criminal prosecutions." Clearview took down the video.

Apple revoked Clearview's enterprise account, which had let the company distribute its app outside of the App Store, because such an account was intended for private distribution of internal workplace apps to employees, not as a way to give a secret superpower to police and potential investors.

Facebook, Venmo, Twitter, LinkedIn, and Google sent Clearview cease-and-desist letters demanding that the facial recognition company stop collecting data from them and delete any photos already in their possession. But none of them took it further than that, perhaps because LinkedIn had just lost its high-profile battle against that data scraper, and a federal court had declared that it was legal to collect public information on the internet.

Lawsuits were filed against the company across the country and Clearview was hammered in the press. Local journalists filed public records requests to see if police in their jurisdictions were using the app. A journalist at HuffPost dug into the company's ties to the far right. I heard that Charles Johnson was involved in the company and reached out to him, a call that would eventually open a box that Clearview had given away 10 percent of its equity to try to keep closed. An anonymous source leaked internal Clearview records to BuzzFeed News that included a list of more than 2,200 organizations that had used the company's app, along with how many searches they had run: nearly five hundred thousand collectively.

What Clearview had done represented a turning point. We were on the cusp of a world in which our faces would be indelibly linked to our online dossiers, a link that would make it impossible to ever escape our pasts. That terrified people, so much so that the norm-violating company seemed on the brink of destruction, its onslaught on privacy too dramatic to ignore.

But then a greater threat appeared and turned the tide away from privacy once again: A pandemic hit.

PART III

FUTURE SHOCK

CHAPTER 17

"WHY THE FUCK AM I HERE?" (2020)

On a Thursday afternoon in January 2020, the day before Hoan Ton-That and I met for the interview at WeWork, Melissa Williams got a call on her cell phone. It was an officer from the Detroit Police Department. He was calling about her husband, Robert Julian-Borchak Williams.

"He needs to turn himself in," said the officer, who seemed to be under the impression that she and Robert weren't together anymore and that she was a "baby mama." Melissa thought that there was racial prejudice at play; Robert was Black, and Melissa was white. They were in fact happily married with two young daughters and lived in a well-off suburb of Detroit called Farmington Hills.

"What is this regarding?" Melissa asked, indignant.

"We can't tell you, but he needs to come turn himself in," the officer responded.

"Why don't you call him?" Melissa said.

"Well, can't you just give him the message?" the officer replied.

Melissa didn't know what to think. She was picking up her five-year-old from kindergarten and did not have time to deal with that nonsense. She said the officer should call her husband directly and provided his cell phone number. Then Melissa, who worked in real estate and did food blogging on the side, hung up and called Robert, who was in his office at an automotive supply company.

"I just got a really weird call," she said to him. "What did you do? What is this about?"

She described the call. Robert's forty-second birthday was two days away, so he thought at first that one of his friends was playing some kind of practical joke. But as he and Melissa talked, he heard the beeping of another call coming in. He hung up with Melissa to take it. It was a police officer. It was no prank.

"You need to come turn yourself in," the officer said.

"For what?" Robert said.

"I can't tell you over the phone."

"Well, I can't turn myself in, then."

The officer didn't like that. "You need to come down before we come out to your job and make a spectacle. You don't want that," he said.

"What is this about?" Robert insisted.

"I can't tell you right now. Just know that it is a serious matter and that it'll be easier if you just come turn yourself in," the officer said, as Robert later recalled the conversation.

"Well, since I don't know what you're talking about and I don't think I did anything wrong, I'm not turning myself in," Robert said. He may also have told the officer to "eat a dick." Then he hung up.

Meanwhile, Melissa had arrived home with their eldest daughter, Julia. The Williams family lived in a split-level home with a nice grassy yard in front and stately trees in the back. Melissa's mom arrived soon after, having picked up their two-year-old, Rosie, from daycare.

As Melissa was talking to her mom and getting Julia and Rosie settled, she looked out her front window and saw a squad car pull in, a black Dodge Charger with DETROIT POLICE stenciled in light gray on its side. Only then did it fully sink in that this was for real.

Two officers got out of the car and came to the door. "Send Robert out," one of them said. The officers must have seen the two cars in the driveway and assumed that the second one was Robert's.

Melissa said he wasn't home yet and again asked why they wanted him; she was getting more panicked. Robert happened to call her just then. He was driving home and wanted to know what he should pick up for dinner. He had no idea how serious the situation had become. "You have visitors here," Melissa said, flustered but trying to stay calm so that her young daughters wouldn't get upset. "There are two cops here from Detroit."

"Where are they?" Robert responded.

"They're in the doorway."

"Get them out of our house," he said. He could hear the officers in the background, asking Melissa where he was and when he was coming home. "Tell them I'll be there when I get there."

Melissa hung up and asked the officers again what it was all about. They said they couldn't tell her and then got back into their squad car, pulled out of the driveway, and drove a few houses down to park and wait. Melissa's mom left as well, sure that it was all a big misunderstanding, easily resolved.

Ten minutes later, Robert pulled his white SUV into the driveway. The squad car pulled up quickly behind him, blocking him in. Two officers, both wearing body cameras that recorded the episode, jumped out and rushed up to Robert before he was even fully out of the car. He shut the door behind him, and the officers got close up to him, so that Robert was trapped between them and his car.

"Are you Robert Williams?" asked a young-looking officer named Mohammed Salem, who stood closest to Robert.

"Yeah," said Robert.

"You're under arrest. We have a warrant for retail fraud in the first degree," Officer Salem said. "Go ahead and—"

"Retail fraud," Robert said, in a confused tone.

The second officer, a taller, beefier guy named Alaa Ali, stepped toward Robert and grabbed his left arm.

"—put your hands behind your back," Officer Salem continued. "I'm not going to drive over here all the way from Detroit—"

"Settle down," interrupted Robert, looking at the two strangers with an impressive calm, as if it were his job to keep the situation from escalating, rather than that of the officers.

"No, no, no, no, no, no," said Officer Ali, the one who had Robert's left arm in his grip. "We're the police. Don't tell us to settle down. This isn't a game. Let's not make a show in front of your wife and kids."

"Why are you grabbing me, bro?" Robert asked, turning his head to look at him.

"Because you are under arrest for your warrant."

"Can I see the warrant?" Robert asked.

"We will show you the warrant in the car," Officer Ali said, forcing Robert's arm behind his back to handcuff him.

Robert seethed. He hadn't done anything wrong. He wasn't a criminal. "This is a gross misunderstanding," he said.

What Robert could not have known as he stood in distress on his front lawn, his wife and daughters watching from the house, was that his case would become the first known account of an American being wrongfully arrested based on a flawed facial recognition match.

Officer Ali was standing behind Robert, trying to handcuff him, but it wasn't easy. Robert Williams was a large man, over six feet tall and weighing more than three hundred pounds. "I'm going to have to three-cuff him," Officer Ali said to his partner, meaning he would need three pairs of handcuffs to trap Robert's hands behind his back.

"I got you," Officer Salem said, handing over his pair.

Once Robert was handcuffed, the officers became noticeably less aggressive. Officer Ali went to the car to get the warrant and brought back a white sheet of paper but repeated that he wouldn't show it to Robert until they were in the squad car.

Melissa was standing in front of the house in her socks, looking resigned and sad. Officer Salem waved her over and told her she should take Robert's phone, wallet, and money, particularly the money, because the detention center would put any cash he had onto a debit card and charge a percentage fee to do so. Melissa gingerly stepped around snow on the driveway. "Where are you taking him?" she asked.

"To the Detroit Detention Center," Officer Salem said, and then he gave the address. "I'll write it down for you."

"If you want to put on shoes, we got time," Officer Ali said, finally seeming a little sympathetic. Melissa turned around silently and went back inside the house.

Robert continued asking why he was being arrested. He gestured at the supposed warrant in Officer Ali's hands. "Can I see what we're looking at here?"

Officer Ali handed the warrant to Officer Salem, who confirmed Robert's full name and date of birth and then held it out for Robert to read. "Felony for shoplifting," Robert read aloud, sounding confused. "Retail fraud, first degree."

"Remember? It happened two years ago," Officer Salem said, as if that clarified things.

"No," said Robert, shaking his head.

"We don't make the cases," Officer Salem volunteered. "The paperwork just comes to our desk and we come to you," Officer Ali added. Officer Salem kept going: "Whether you're a murderer or a rapist, it doesn't matter."

Melissa had come back out of the house, wearing shoes and holding two-year-old Rosie, who overheard this.

"Hi, Rosie butt," Robert said, trying to force cheerfulness into his voice. Melissa reached into Robert's pocket to take his things. Five-year-old Julia came out of the house and stood behind her mother.

"Hey, Juju bean, where's your jacket?" Robert asked, and then he saw the look on her face. She looked scared. "They made a mistake. It's okay. I'll be back in a minute. Go in the house. Do your homework."

Robert was dazed, in disbelief that this was happening. He tried to stay as calm as possible. He knew how quickly a wrongful arrest could turn deadly. Unarmed men were shot, tased, or choked to death by police. Robert did not want to become a statistic in the police brutality debate.

The officers walked Robert to the squad car, shifting the passenger seat up to squeeze him into the back. Melissa asked again where he was being taken.

"Google *Detroit Detention Center,*" Officer Ali said.

"You're taking my husband in front of my children for something that . . ." Melissa said, trailing off as Officer Ali interrupted her.

"I understand," he said. "But you saw the warrant."

"I didn't see the warrant," she said, upset.

Officer Ali walked to the car and climbed into the passenger seat.

"Have a better night," Officer Salem called to Melissa as he sat down behind the steering wheel.

Then the officers drove off with Robert in the back of their car.

None of the neighbors had come outside while it was happening, but it was 6:00 P.M. by then. People were home from work; they had surely been at their windows, watching and wondering what Robert had done.

Melissa went into the house with her girls, and they started sobbing. "Where did Daddy go?" Julia asked.

The detention center, a campus of redbrick buildings surrounded by a barbed-wire fence, is on the east side of Detroit. The officers made small talk in the car during the forty-minute ride there.

"You have a nice car," they said. "You have a nice house. That's a quiet neighborhood."

Yeah, yeah, yeah, Robert thought. The only thing he wanted to hear was what he was accused of doing.

Robert's arrival at the detention center was a bit of a blur. The people there took his belt, the laces from his shoes, and the drawstring from his jacket—anything he could use to hang himself. He was frisked and patted down. He posed for mug shots and had his fingerprints taken, and then he was put into a holding cell with ten other guys.

"What are you in for?" one of the guys asked.

"I don't know," Robert said. "Shoplifting, apparently."

"What'd you steal?"

"I didn't steal anything."

Around 9:30 P.M., Robert was allowed to make a phone call to Melissa.

"So what's going on?" she asked.

He still didn't know. They both racked their brains. *What could this be about?*

Robert didn't know how long he was going to be stuck there. He asked Melissa to email his boss if he wasn't home by morning to say he had a family emergency. The absence would break his perfect attendance record.

Around 2:00 A.M., officers came to the cell and called Robert's name. He grabbed his jacket, elated. He thought he was being released. But they were there to swab his mouth and get a DNA sample.

Robert spent a long, sleepless night on the cell's concrete bench. He went through all his friends in his head; could one of them have gotten into trouble, and given the police Robert's name? How the hell had he wound up here? He wanted to be home, sleeping in his own bed next to his wife. Or on his couch. Or hell, even in a dining room chair. Anything would be better than here.

The next morning, he was put in handcuffs again and taken to an interrogation room. There was a table and three chairs. Robert sat down and two white detectives sat down across from him. One of them gave him a piece of paper explaining his rights and asked Robert to read it aloud. While Robert read, the other detective pulled a stack of papers out of a folder.

The first detective asked if Robert understood his rights and then asked whether he wanted to make a statement. "I don't even know why I'm in here," Robert replied. The detectives said they couldn't tell him more unless Robert signed the sheet of paper in front of him, waiving his Miranda rights and agreeing to be questioned without an attorney present. This was not true. They could have told him what was going on, but it was an effective way to get him to agree to be interrogated. It was as if the police department had used Franz Kafka's *The Trial* as a training manual, aiming to keep Robert in a state of confusion for as long as possible. It was absurd: How could the police arrest him, hold him overnight, and keep refusing to tell him why?

After Robert signed, the detective asked whether Robert wanted to ask a question before they began. "Yes," he said. "Why am I here?"

They explained that they were investigating multiple larcenies at retail stores in downtown Detroit. Then the interrogation began.

One of the detectives took a piece of paper from the stack and set it down in front of Robert. It was a still image from a surveillance camera of a heavyset Black man.

"Can you do me a favor?" the detective said. "Is that you?"

"No," Robert said, almost laughing. He held the image up next to his face, so the detectives could see for themselves. "Not even close."

He slapped the paper back down on the table. "I'm pissed," he said. "Keep going."

The detective took another paper from the stack. It was another surveillance camera still, of what looked like the same man but dressed in different clothes, wearing a red St. Louis Cardinals cap. He was standing in a store. "Is that you?" the detective asked.

Robert didn't think he looked anything like the guy. Being large and Black was all they had in common. Robert wasn't even a Cardinals fan.

"No," Robert said, emphatically.

"That's not you?" the detective repeated.

"At all," Robert replied, shaking his head.

The detective put a third piece of paper in front of him, a close-up image of the man's face.

"Not there either," said Robert. "Y'all can't tell that?"

The detectives stared at him and then shuffled their papers in awkward silence.

"So on one of the pictures, we actually got facial recognition," the detective said. Robert nodded knowingly.

The detective grabbed a fourth sheet of paper, an "Investigative Lead Report," and showed it to Robert. It had the image of the man in the Cardinals cap and, next to it, Robert's photo from his last driver's license, a gray scarf around his neck, his broad head cocked, his beard carefully manicured, and his eyebrows raised so that he had a slightly goofy expression.

"That's not me," he said pointing at the man in the hat. Then he pointed at the photo of himself with his head cocked. "That's me."

Robert had no idea that getting a driver's license meant having his photo included in a perpetual facial recognition lineup.

The detective asked if he'd ever been to Shinola, an upscale boutique that sold watches, bicycles, and leather goods in a trendy Detroit neighborhood called Midtown.

"In 2014, in the summertime, when they first opened," Robert said. He remembered the year because he and his wife had the newborn Julia with them. "And we haven't been back."

One of the detectives got a resigned look on his face, but they kept questioning him, asking whether he'd been to John Varvatos, a high-end male clothing shop. Robert said he had never been there. The tone of the conversation changed somewhat. One of the detectives, at least, seemed to be convinced he wasn't their guy.

Robert wore a Breitling watch. "My watch costs eight times what a Shinola costs," he said. One of the detectives asked whether he'd ever pawned any watches. Robert said he had pawned a Breitling once, years earlier. The detective then admitted that they already knew that. But what had once seemed like incriminating evidence now looked like a guy who hadn't wanted a watch he bought. The detective who seemed convinced of Robert's innocence looked at his partner with a glum expression on his face, then turned back to Robert. "You got any questions?" he asked.

"Yes," Robert replied. "Why the fuck am I here?"

ROBERT WILLIAMS WAS arrested two days before his birthday because a man had walked into a Shinola store in October 2018 to steal some watches and the theft had been recorded.

The heavyset man in the Cardinals cap entered the Shinola a little after 5:00 P.M. on a Tuesday afternoon. According to a police report, a Shinola employee greeted Not-Robert, who was wearing a black jacket and carrying a pink T-Mobile shopping bag. Not-Robert set his pink bag down on a table near the entrance and then walked to the men's section at the back of the store.

A display shelf blocked the man from the employee's sight, but a surveillance camera made by the China-based Hikvision captured what happened next. Not-Robert looked around and, content that no one was watching him, grabbed five watches worth $3,800 from a display and stuffed them into his jacket. Then he walked over to the table to get his pink bag and walked out of the store. He had been there for only a few minutes.

An employee soon noticed the missing merchandise and reported it to a loss prevention firm that Shinola relied on for incidents like that. A week later, an investigator at the firm reviewed the day's surveillance video, found the footage of the theft, and sent a copy to the Detroit police.

Five months passed, the case seemingly at a standstill. The surveillance camera had been near the store's ceiling, capturing the shoplifter from above, but his forehead had been covered by the hat. It wasn't a perfect capture of the face, but it was the only lead.

Michigan had a facial recognition database with almost 50 million photos, including mug shots, driver's licenses, and state IDs. Michigan had two different facial recognition algorithms to search those photos; both were top performers in NIST's Facial Recognition Vendor Tests.

The year prior, NIST had looked at algorithmic bias, or "demographic effects," as the agency put it in its report. The test had limitations: It had been done using relatively high quality government images—mug shots and visa application headshots. The algorithms hadn't been tested with the kinds of images most likely to be used by police in an investigation, such as the still from Shinola's surveillance camera. Even so, NIST found clear problems; in the nearly two hundred algorithms they tested, there was a wild variation in performance across a number of factors, including race. The media takeaway was that facial recognition technology was biased, but the findings were more nuanced than that; there were good algorithms and there were bad algorithms, as the expert Clare Garvie put it.

NIST had looked at two types of facial recognition: a 1:1 search, comparing two faces to determine whether they were the same person, and a 1:*n* search, looking for a face in a big database full of many people. The worst bias problems happened with the unlock-a-smartphone type facial recognition. If the subject was female, Black, or Asian, smartphones using these algorithms were more likely to lock them out or to let someone who looked like them in.

This usually happens because the company that made the algorithm didn't train its neural network well enough: It gave it only photos of middle-aged white men, for example. The resulting algorithm might be able to discern among white men very well, but it wouldn't be as good at seeing women, Black people, Asian people, or senior citizens. That had historically been a problem with computer vision, but after years of embarrassing, offensive mistakes—web cameras that couldn't detect a Black face (2009), image recognition that saw Black people as gorillas (2015), soap dispensers that worked for white hands but not Black hands (2017)—most of the engineers behind artificially intelligent products finally started to get the message. They needed to train their software with diverse data sets to work on diverse people.* But it

* The mistakes didn't stop overnight. As recently as 2021, a user watching a news story on Facebook about a group of Black men got an AI-powered prompt to "keep

wasn't just a matter of *who* was included; the physics of photography itself might also be an issue.

Some experts questioned the underlying technology of cameras—that they may be designed to best capture lighter skin. There was a predigital precedent; in the 1950s, Kodak had distributed a "Shirley card" to photo labs to calibrate printers to produce the best version of the photo of the woman on the card. The models for Shirley were initially white women. Both the film and the printer were optimized for a person with lighter skin tones, leaving others under- or overexposed in photos. It wasn't until the 1970s that Kodak realized that it had a problem, changing its film to capture a wider range of dark colors and diversifying the women on its Shirley cards. Problems persist; in dimly lit rooms in the twenty-first century, photographers and film directors still have to light Black subjects more intentionally to make sure their faces are captured well by the camera.

In NIST's test of big database searches, bad algorithms had bias problems, particularly when it came to Black women. But the most accurate algorithms showed virtually no bias in testing. They worked equally well on all demographics, with overall accuracy rates approaching 100 percent under good conditions and over 90 percent in more challenging scenarios.

The algorithms weren't inherently biased; their performance became so based on the data they were fed. Realizing that, companies went to extreme lengths to acquire more diverse data sets. Chinese companies that made deals to install facial recognition systems in African countries said they needed the diverse faces to improve their technology. In 2019, Google asked a contractor to collect real-world face scans from people with dark skin, and the contractor reportedly targeted homeless people and university students, giving them $5 gift certificates in compensation. Google executives later said they were disturbed by the contractor's tactics and that the company had simply been trying to ensure its Face Unlock feature for Android "works across different skin tones and face shapes."

---

seeing videos about Primates." Facebook apologized and disabled the recommendation feature.

The algorithmic fruits of these types of exercises topped NIST's charts. The two highly rated algorithms that Michigan was using—one from the Japanese tech giant NEC, the creator of the celebrity doppelgänger exhibit at the 1970 Osaka World Exposition, and another from a newer company based in Colorado called Rank One Computing—were supposed to work well regardless of a suspect's race, age, or gender. But even a highly accurate algorithm, deployed in a society with inequalities and structural racism, will result in racist outcomes.

A DIGITAL IMAGE examiner for the Michigan State Police named Jennifer Coulson ran the search on the Shinola shoplifter's face. A former parole agent, Coulson had special training in running facial recognition searches. Courses offered by the FBI as well as facial recognition vendors went over the limitations of the technology, the importance that search images be high quality, and the clues to consider in a potential match to confirm the likelihood that it was the same person. Critics say, however, that the supposed training too often focuses on how to use the software, not the incredibly difficult human task of comparing a face to a selection of doppelgängers presented by a computer, a lineup that may well not include the person sought.

After Coulson submitted the still of the Shinola shoplifter, what's called a probe image, the facial recognition system spat out its best guesses: 243 photos ranked in order of likelihood to be the same guy. Williams's goofy driver's license photo was number nine on the list, and Coulson decided it was the best of the options.

In March 2019, Coulson generated that investigative lead report for the Detroit police. It included the probe image next to Robert's driver's license photo, along with his name, birth date, and license number. Above the photos was the message required at the top of all facial recognition reports: "THIS DOCUMENT IS NOT A POSITIVE IDENTIFICATION. IT IS AN INVESTIGATIVE LEAD ONLY AND IS NOT PROBABLE CAUSE TO ARREST. FURTHER INVESTIGATION IS NEEDED TO DEVELOP PROBABLE CAUSE TO ARREST."

This is what technology providers and law enforcement officers always emphasize when justifying the use of facial recognition technology: It is supposed to be only a clue in the case, a lead, not a smoking gun. But that ignores the known phenomenon of confirmation bias. Generally, once officers have a suspect, they start to assume that person is guilty and look favorably on anything that corroborates that belief, even if it's weak, such as the fact that a man accused of stealing watches once pawned one.

Once Coulson and the algorithm had pointed the finger at Williams, investigators were supposed to go out searching for independent evidence—location data from his phone or proof that he owned the clothing that the suspect was wearing—but they do not appear to have done anything so involved. Instead, they just looked for another person who thought that Robert Williams looked like the guy who had walked into the Shinola that day.

At least five months after the theft, the police reached out to Shinola, saying they wanted an employee who had been at the store when the crime happened to look at a photo lineup. An employee who was there that day declined to do so because it had been months and they couldn't remember what the man had looked like.

So the investigators went to the woman at Shinola's loss prevention firm who had reviewed the surveillance footage and sent police the clip of the shoplifting incident. They showed her photos of six Black men, including Robert. She hadn't been there the day the theft occurred. She had never met the guy in the Cardinals hat or seen him in person. She had only watched the theft on her computer screen. She had a still from the surveillance video on her phone; she looked at it, looked at the photo lineup, and then agreed with Coulson and the computer that Robert Williams looked the most like the guy in the video. That woman, like Coulson, was white, which is significant because researchers have found that people are worse at identifying the face of a person of a different race. Law enforcement officers, though, decided that that was sufficient evidence to charge him.

That was why Robert was arrested. It was not a simple matter of an algorithm making a mistake; it was a series of human beings making bad decisions, aided by fallible technology.

AFTER THE DETECTIVES finished their interrogation, they offered Robert water, put him back into handcuffs, and took him back to his holding cell. It turned out they weren't the ones who had issued the warrant for his arrest; they were working a related larceny with the same suspect. They said they couldn't intervene in his case. He had to appear in court later that day and formally plead not guilty. Only after that was he released on a $1,000 bond, the amount set relatively low because he didn't have a criminal history.

Robert got out of the detention center around 10:00 P.M. Friday, the night before his birthday. He waited in the rain for Melissa to come pick him up.

But the nightmare wasn't over. He was charged with a felony. The Williams family contacted defense attorneys, who quoted prices of at least $5,000 to represent him.

"I'm innocent and they have no proof whatsoever," Robert said.

"You're charged with a felony, I have to charge a felony price," one replied.

Melissa tweeted at someone who worked for the American Civil Liberties Union of Michigan. "I'd love to talk to you about Detroit's facial recognition & my family," she wrote.

"Email me," the ACLU employee replied. Soon the organization was helping the couple find a lawyer.

Using Instagram, Robert was able to figure out what he had been doing on that Tuesday evening in October 2018 when the shoplifting had occurred. He had been driving home from work and had posted a video to his private account when a song he loved came on: 1983's "We Are One" by Maze and Frankie Beverly, a jazzy number about treating one another with more humanity. He had an alibi, had the Detroit police bothered to check.

Two weeks after his arrest, Robert took a vacation day to appear in a Wayne County court. With the thin evidence against him, the prosecutor asked the judge to dismiss the case, but "without prejudice," meaning that Robert could later be charged for the same crime.

Months later, the ACLU introduced me to Robert and Melissa. I interviewed them, investigated what had happened, and wrote a front-page story in *The New York Times* titled "Wrongfully Accused by an Algorithm."

In response, the prosecutor's office issued an overdue apology and a promise to expunge Robert's fingerprint data. "This does not in any way make up for the hours that Mr. Williams spent in jail," the prosecutor's office said.

The Detroit police chief admitted that the police work in Robert's case had been "sloppy" but said that he still saw facial recognition technology as a valuable investigative tool. The city had decided not to use facial recognition for minor offenses, such as shoplifting. It would use it only for violent crimes and home intrusions—the sorts of crimes for which investigators would, in theory, gather evidence more diligently.

But no one called Robert and Melissa to say sorry. The stress of what had happened wore on him. At the end of 2020, he had a series of strokes. For a while, half of his face was paralyzed. In early 2021, Robert and Melissa sued the city for wrongful arrest, a case that remains ongoing as of the writing of this book.

Soon after Robert's story came out, a second case of misidentification surfaced in Detroit. A thief had been recorded snatching a smartphone, and facial recognition led a detective, the same one who had worked Robert's case, to charge and arrest a man named Michael Oliver. But Oliver was clearly not the same person. His arms were heavily tattooed, while the dark-skinned arms of the person who had stolen the phone were bare.

In a third case, a man in New Jersey spent more than a week in jail after a bad facial recognition match. There was a fourth man in Maryland. And then a fifth: A twenty-eight-year-old man named Randal Reid was arrested in Georgia, and held in jail for a week, for buying designer purses with stolen credit cards at consignment stores in Louisiana. But Reid had never been to Louisiana and was at work five hundred miles away when the crimes happened. A Jefferson Parish detective had used facial recognition technology to identify a man from a surveillance camera still, who happened to look a lot like Reid. The facial recognition vendor the detective's agency contracted with, for $25,000 a year, was Clearview AI.

In every case, the man wrongfully arrested was Black.

More than a year after Robert's arrest, it still haunted their oldest daughter, Julia. She would bring it up at the oddest times. One summer afternoon while Robert and Melissa were weeding their yard, Julia ushered everyone inside for a family meeting. She closed the curtains and said it was very important that they find the man who had stolen those watches at the Shinola store. She sketched her version of his face.

"I don't know when this is going to go away," Robert said.

CHAPTER 18

A DIFFERENT REASON TO WEAR A MASK

At the beginning of January 2020, the same month that Clearview's existence became public, the World Health Organization (WHO) reported a "cluster of pneumonia cases" in Wuhan, China. By the middle of January, scientists had identified the culprit, a virus they had never seen before, which they named SARS-CoV-2. The pneumonia-like disease it caused was named Covid-19. Within a day, the first case of Covid-19 outside China was recorded, in Thailand. By the end of January, there were nearly eight thousand cases worldwide and the WHO declared a state of emergency; it was an outbreak.

In the months that followed, the virus spread around the world with alarming speed. Case numbers climbed. Many of the infected lost their sense of taste and smell. Some couldn't breathe and were put on ventilators. Hospital beds filled. Seemingly healthy people contracted the virus and were dead within a month. Many survived it but had long-term symptoms: lingering coughs, heart murmurs, neurological conditions.

It was a scary time, in part because of how little the world knew about the disease. At first people wore gloves and disinfected their groceries, convinced that the virus was spread through physical contact. But in fact, it was in the air around us, exhaled from one set of lungs and breathed in by another.

Governments ordered lockdowns, asking everyone to stay home to try to halt the disease's inexorable spread. The global economy came to a bone-jarring halt. Hundreds of millions of people contracted the virus, and millions died of it. People's faces disappeared behind masks, along with the uproar about how they were being tracked. A pandemic took priority over privacy.

Clearview AI, as if disappointed to fall out of the news cycle, sought to capitalize on the crisis. In March 2020, *The Wall Street Journal* reported that the company was "in discussions" with unnamed state agencies "about using its technology to track patients infected by the coronavirus." The unverified claim sounded like startup hype, but it was true that the company's software could theoretically be used for tracking people with Covid-19 and anyone they came into contact with.

Reports emerged that South Korea, China, and Russia were tracking faces to try to control the spread of the virus. After traveling abroad, one Moscow resident was ordered housebound for two weeks; when a few days passed without symptoms, he decided to get some fresh air by going outside his building to take out the trash. Thirty minutes later, police showed up at his door. News reports suggested that he'd been spotted by a facial recognition camera.

Were U.S. policy makers also contemplating that level of surveillance of Americans? Were they really considering hiring Clearview to do it?

"We can't let the need for COVID contact tracing be used as cover by companies like Clearview to build shadowy surveillance networks," Senator Edward Markey tweeted at the end of April 2020 in response to the *Journal*'s story. "I'm concerned that if Clearview becomes involved in our response to #COVID19, their invasive tech will become normalized, and that could spell the end of our ability to move anonymously in public."

A state representative in California, Democrat Ed Chau, tried to pass a bill that would have cleared the way for private companies and government agencies to use facial recognition technology to fight Covid-19. But that bill didn't gain traction. No public contracts emerged indicating that Clearview was involved in pandemic-related work. What Clearview AI and many other facial recognition technology companies did instead during that time was to adapt their facial recognition systems, training them on images of people in surgical and

N95 masks to ensure that their software would still be able to recognize those who cared to take Covid-19 precautions.

The pressure on Clearview eased somewhat. All the media attention had caused controversy, but it had also made more law enforcement agencies aware of the company's existence, resulting in new trials and new customers. Clearview tried to reassure the public that its tool would be used only in the service of justice, shutting down the accounts of users who weren't in law enforcement and publishing a "code of conduct" on its website: "Clearview AI's search engine is available only for law enforcement agencies and select security professionals to use as an investigative tool, and its results contain only public information."

But in the months after the story came out, as the company sought to raise another $20 million from investors, Richard Schwartz privately circulated a slideshow suggesting that Clearview was still trying to sell its product to businesses. It had booked $1 million in revenue in 2019, according to the presentation, and was growing—new daily sign-ups had doubled since the first *New York Times* story came out, and the daily number of searches was up from 1,384 to 2,408. Real estate firms and retailers were using a Clearview product called Insight Camera that identified people who walked into a building, according to the deck, and Clearview planned to provide its software to the "sharing economy and other commercial users" as a background-checking tool.

The presentation had a promotional pull quote from my *Times* exposé, though it hadn't been intended to be flattering: "[Clearview is] far beyond anything constructed by the United States government or Silicon Valley giants."

One of the companies that had tried out Clearview's tool early on was Madison Square Garden, an iconic arena in New York City, the home stadium for the basketball team the Knicks and the hockey team the Rangers, where famous musicians and comedians perform. Madison Square Garden wound up adopting facial recognition technology, from a different vendor, in 2018, ostensibly to turn away known rabble-rousers and security threats from its venues. But eventually, the venue used the technology as a punitive measure.

In 2022, the Garden put thousands of lawyers on its face watch list because their firms had sued either the entertainment giant or one of the many restaurants and event venues it owned. As the barred lawyers tried to get into concerts, shows, and games at the arena, they passed through metal detectors where cameras scanned their faces and matched them to photos gathered from their firms' own websites. Once they were matched, security guards would pull them aside, confirm their identities, and inform them that they weren't welcome until the cases their firms had brought were resolved. That happened to lawyers who had no involvement in the cases but happened to be employed by the firms in question.

The ban could not have been effectively enforced without facial recognition technology. Most of our antidiscrimination laws are predicated on categories that are typically visible at a glance: race, gender, disability. Technology such as Clearview's will make possible a new era of discrimination, allowing businesses to turn away customers based on what is knowable about them from an internet search of their faces. Businesses could use the tech to bar anyone who left them a negative review or ran afoul of their owners' political beliefs. Some stores might turn away antivaxxers, others radical leftists.

Prospective investors were spooked by the backlash against Clearview, and the company struggled to raise new funds. The existing investor David Scalzo, meanwhile, was frustrated with the company's timid public messaging. He had invested in Clearview because it was going to change the world and be as universal as Google—and the ubiquity would translate into a bonanza for early investors like himself. But now Clearview was painting itself into a corner, saying it would work only with police. "They're limiting the eyeballs. That's not a business model," said Scalzo, who wasn't going to make the returns he wanted on a glorified government contractor.

Ton-That and Schwartz told him that that was strategic. Characterizing it as a tool explicitly for police would help the company win lawsuits, given the country's reverence for crime fighting. That did not reassure Scalzo. The only reason Clearview was leading the field was because potential competitors were scared. Winning the lawsuits would clear the way for bigger technology giants to rush in and take over. "The delta between the path they're on and the path they could be on is a trillion dollars," Scalzo said. "Let it be free! Instead they're artificially caging it. If this thing went to consumers, it would go to fifty million users in, like, four days."

He had thought that Ton-That and Schwartz had spines like that of Travis Kalanick, the Uber chief executive who had rolled his ride-hailing app out around the world, ignoring lawsuits and regulators screaming at him that the business model wasn't legal. Scalzo said that Clearview had transformed from a rocket ship into a "GoFundMe for lawyers."

That wasn't all. It was also turning into a punching bag for global privacy regulators.

CHAPTER 19

I HAVE A COMPLAINT

In January 2020, Matthias Marx, thirty-one, was deeply troubled after reading the *New York Times* article about Clearview AI. He was a technologist who cared deeply about online privacy. As a PhD student at the University of Hamburg, he built tools to help people avoid inadvertently sprinkling their data around the internet. He was also a member of the Chaos Computer Club, a hacker collective obsessed with protecting digital rights.

Though Marx considered himself knowledgeable about the different ways people's digital movements are tracked, he hadn't thought much about the face being one of them. Reading about Clearview AI, he wondered if the American company, which claimed to have amassed a database of billions of faces, had his.

So he sent the company an email. "Dear sir or madam," he wrote. "Please provide access to my personal data." He included a digital headshot. He was standing outside wearing a black T-shirt, in strong contrast to his pale, freckled skin. He had squinty blue eyes, a generous nose, and a broad neck. When the photo was taken, he was smiling slyly, his longish blond hair blowing in the wind.

That might seem like an oddball request, a message that the company would ignore or delete. Why would New York–based Clearview respond to some random grad student from Germany asking for information about himself? Clearview had data on millions of people, after all. How could it possibly take the time to answer emails like that from any of them?

Well, Clearview didn't have a choice. European law required the company to respond to Marx and to do so within thirty days. If Clearview wanted to sell its facial recognition tool in Europe—which it was trying to do at the time—it had to respect one of the fundamental rights of EU citizens: "access to data which has been collected concerning him or her, and the right to have it rectified." The right was reinforced and strengthened in 2018, when the European Union put into effect the world's most stringent privacy law with the world's most boring name: the General Data Protection Regulation, or GDPR. Marx not only had the right to ask for any photos Clearview AI had of him; he had the right to demand that the company delete them.

But Marx did not think it would come to that. In fact, the request was a bit of a lark. He was a private person and avoided putting photos of himself on the public internet, so it seemed unlikely that the American company would have him in its database, but it was worth a try.

A week later, he got a response from Clearview, addressed erroneously to "Christian," asking for a government-issued ID. "We need to confirm your identity to guard against fraudulent access requests," said the email from privacy-requests@clearview.ai. Clearview needed to ensure that Marx was Marx, and not an impostor trying to get his data.

Clearview also wanted the ID to see where Marx lived. "These rights are not absolute," the email said, noting that they "vary by jurisdiction." If Marx truly lived in the European Union, he could get his data, but if he lived in a part of the world without a "right to access" law, Clearview would not comply with his request. Companies aren't generally in the business of providing rights lawmakers haven't guaranteed.

Marx ignored the email. Regardless, a month later, Clearview sent him another, with the subject line "Data Access Request Complete."

Marx opened the PDF attached to the email and stared at the screen in surprise. The report included the headshot Marx had sent to Clearview, the one in the black T-shirt, which the company had used to conduct its search. Next to the headshot were the results: two photos of Marx taken eight years earlier, when he was in college, his face a little thinner and his hair a little shorter. He was wearing a shirt that said, HELL YEAH IT'S ROCKET SCIENCE.

The report included links to the website where the photos had appeared, a British stock photo site called Alamy. Marx wasn't familiar with it, but he recognized the photos. They had been taken for a newspaper story about a project he and some other students had submitted to the $20 million Google Lunar XPRIZE—a wild race to land a craft on the moon, a competition that no one ultimately won. The newspaper, or the photojournalist it had hired, must have sold the photos to the stock image site at some point.

"Oh, krass, damit habe ich nicht gerechnet," Marx tweeted along with a screenshot of the report. (Translation: "Oh, wow, I didn't expect that.")

Marx wrote a complaint to his local privacy agency, the Hamburg Data Protection Authority (DPA). He said that Clearview was processing his data without his consent and that the company hadn't fully complied with his access request. It had sent him only photos and not the facial recognition template that had been created for him. This might fly in America, where privacy laws barely existed, but it seemed clear to Marx that what Clearview had done was illegal in Germany.

A couple months later, out of the blue, Clearview sent Marx another report. It contained the same two images as well as eight others—of people who looked similar to Marx. They were all profile photos from VK, Russia's version of Facebook, except for one, which came from an Instagram clone site. Had Clearview expanded its database, he wondered, or changed its search to return a greater range of results? Either way, he didn't care. Marx was convinced that the investigative tool, so popular with police, was acting illegally in Europe. He forwarded the email to Hamburg's DPA, which ruled that Clearview had violated the law and ordered it to delete Marx's biometric information.

But it was an ineffectual order. Eventually, with Clearview scraping all the time, his face would just get collected again. Marx couldn't keep himself out of the database, he could only keep his face from coming up in search results, and he had to give Clearview permission to use his face in order to do that. He did not consider this an acceptable outcome but there was nothing more he could do.

CLEARVIEW'S GROWING TEAM of lawyers had their hands full. Clearview AI had been making inroads globally, and that was now translating into legal inquiries from around the world.

The app had been tried out by security forces in dozens of countries—Belgium, Brazil, Denmark, Finland, India, Ireland, Latvia, Lithuania, Malta, the Netherlands, Norway, Portugal, Saudi Arabia, Serbia, Slovenia, Spain, Sweden, Switzerland, and the United Arab Emirates, among others—and the startup would have to contend with investigations by privacy regulators in Australia, Canada, France, Greece, Italy, and the United Kingdom. The international backlash caused the company to pull back. Hoan Ton-That started telling journalists that Clearview had decided to sell its product, "for now," only to law enforcement agencies in the United States.

After investigations that lasted a year or more, the cohort of international privacy regulators all came to the same conclusion: Clearview had violated their data protection laws. "What Clearview does is mass surveillance, and it is illegal," said Canadian privacy commissioner Daniel Therrien, the first of the regulators to tell Clearview that it could not legally operate in its country.

By the time the regulators were done, Clearview had been declared illegal in at least six countries and was subject to approximately $70 million in fines, which was more than the company had raised from investors and made in revenue. They all declared that Clearview needed consent from their citizens to do what it was doing. Each country ordered Clearview AI to stop collecting its citizens' faceprints and to delete the ones it had. Each time, Ton-That issued a public statement that mentioned how the rulings hurt his feelings, particularly the one from Australia, his native land.

"I grew up in Australia before moving to San Francisco at age 19 to pursue my career and create consequential crime fighting facial recognition technology known the world over," he said in the statement. "I respect the time and effort that the Australian officials spent evaluating aspects of the technology I built. But I am disheartened by the misinterpretation of its value to society."

Italy was one of the countries that fined Clearview and said that what the company had done was illegal. So it was surprising, about a year later, when Italian authorities announced how they had caught a mobster wanted for murder who had been on the lam for sixteen years. The police had run his photo through an unnamed face search engine, and it had unearthed an aged version of him holding a loaf of bread in an article about a pizzeria in France, where he worked the night shift. It's safe to assume he didn't provide his consent to whichever search engine had indexed his photo and made his face searchable.

Ton-That shrugged off the regulators' rulings. He remained optimistic that they would be overruled after the company appealed their decisions. He kept comparing Clearview to Google; they were both search engines, he said, that took information that was already public and made it easier to find.

But Clearview was doing it with a dramatically novel search term: the mathematical map of a person's face. It could do so because faces are so easy to get; people post photos of theirs to the internet indiscriminately. But if Clearview's logic prevailed, it would pave the way for other biometric search engines that relied on what people had put online without realizing it could one day be mined. There could be a voice search engine: Upload a recording of someone talking, and the voiceprint could lead you to any other video or audio of them on the internet. There could be a Gait Google: Upload a photo of a person walking, and gait analysis could lead you to any online video of them moving through the world. Disney World had already explored a primitive form of this. It had a patent on "foot recognition" that would allow it to track guests around its amusement parks with cameras trained on their distinctive footwear.

There was no end to the ways in which we could be trackable in the future without basic rights over the use of our biometric information.

TIMO GROSSENBACHER, A thirty-two-year-old Swiss data journalist, had an entirely different reaction than Marx to the *New York Times* story about Clearview. He was utterly unsurprised.

Anyone could do this, he thought. So he and a colleague gave it a try. They chose a free, prebuilt scraper called Instaloader to download photos; it required some knowledge of the programming language Python, which they had. They could give it a hashtag—they chose ones linked to big events in Switzerland—and then it would find relevant photos on Instagram and download them. Over just ten days or so, they built a database of over 230,000 photos from popular events, including a ten-mile race in Bern and an agricultural fair in Olma. The two men had scraped all the Instagram pictures using just one computer in one location, meaning that a single IP address had requested hundreds of thousands of photos from Facebook-owned Instagram and the company had barely batted an eye, blocking the computer from accessing Instagram only occasionally, for about an hour, before letting it continue to draw the data down.

They also scraped the big kahuna, Facebook, though that was a little more onerous. Their scrapers needed to be logged in to the social network to access photos, so they created ten fake accounts with profile photos from the website This Person Does Not Exist. Then they sent those existential nightmares out cruising on Facebook, downloading the photos of their friends and then their friends' friends, and so on and so forth. The accounts would be blocked temporarily, but would eventually start working again. Over a few weeks, they were able to download seventy thousand Facebook photos.

Then the Swiss duo turned to a freely available facial recognition package called dlib, which Terence Liu and Hoan Ton-That had used in the early days of Clearview to create their celebrity look-alike tool. Grossenbacher used the software to look through the event photos for the faces of 2,500 Swiss politicians, figuring that was ethical because they were public figures.

And *voilà,* the software identified political candidates in hundreds of photos, including a young lawyer who had attended a street parade half naked four years earlier when she had been a teenager. Grossenbacher got the shivers. "Oh, my God," he thought to himself. "What have we done?" (The Swiss data commissioner told them it was an interesting experiment, but that they should never do something like it again.)

It was a shocking demonstration. The building blocks for a face search engine were just too easy to come by: online photos, more and more of them every day, just waiting to be scraped. The facial recognition algorithms to sort through them were increasingly becoming a plug-and-play technology. It meant that even if Clearview were smacked down, copycats would be easy to create. The guys at Clearview "aren't rocket scientists," Grossenbacher scoffed.

Others could build it—and they had—and they were offering it to anyone who wanted it, no matter how sleazy their intentions.

CHAPTER 20

THE DARKEST IMPULSES

David had an addiction.* Ever since he was a kid, he had spent countless hours of his life on the internet searching for naked photos of women and videos of them having sex. He wasn't happy about how much time he spent looking for online pornography, but he couldn't stop.

At some point, he began to want more. He wanted to know who the women really were, to dig beyond their pseudonyms, find out their actual names, and see what they were like in real life. He heard about a tool to do that on a porn discussion board called FreeOnes. Someone had shared a photo of a porn actress, out of character, clothed and smiling, in a news article. Asked how he had found it, the person posted a link to PimEyes, a free facial recognition site.

David, who is in his thirties, checked out PimEyes. It was pretty basic. The landing page said FACE SEARCH in bold white letters, with an invitation to "upload photos of **one** person." PimEyes claimed to have hundreds of millions of faces from 150 million websites. All he had to do was upload a screenshot of a woman and the site would return photos of people it deemed most similar to her. "We do not save your search images," the site promised, adding with painful irony, "Online privacy is very important for us."

The PimEyes face search tool worked. David was able to upload

* David is identified here by a pseudonym. He reached out after seeing my coverage of facial recognition technology in *The New York Times.* He said he had never admitted this to anyone before. He thought it was important that people know what is possible with the technology but was also horrified by his own choices and did not want them linked to his real name.

screenshots of women whose pornography he had watched and get photos of them from elsewhere on the web, a trail that sometimes led him to their legal names. From there, he could know where they lived and find them in the real world, a scary possibility if he had the desire to hurt or assault them. But that was not the appeal for him. Unmasking them was all he sought to do.

"You find them on Facebook and see their personal pictures or whatever and it makes it more exciting," David said. "It's like the secret identity of Batman or Superman. You're not supposed to know who this person is, they didn't want you to know, and somehow you found out."

The women keep their identities secret for a reason. Polite society doesn't tend to approve of their line of work. In 2014, a freshman at Duke University was revealed by a classmate to be a porn actress named Belle Knox. When people found out that the student was paying her $60,000 tuition by having sex on film, she was shunned on campus and harassed online. "I was called a 'slut who needs to learn the consequences of her actions,' a 'huge fucking whore,' and, perhaps the most offensive, 'a little girl who does not understand her actions,'" she wrote at the time.

Ela Darling, a former librarian turned adult film star, got into the industry after an ex-boyfriend threatened to put compromising photos and videos of her online. She decided she'd rather do that herself, and make money off of it. Darling had two Facebook accounts, one for her porn identity under the stage name "Ela Darling," and another associated with her legal name and "vanilla" life. After she friended someone with her vanilla account, she would sign in to her porn account and block the person so that they couldn't see or access that version of herself.* She did that to protect herself, her family, and her friends from

* Darling was also motivated to do this because of "People You May Know," a Facebook feature to help social media users find new connections. Sex workers have complained that their real-name accounts are sometimes suggested as a connection to clients with whom they use pseudonyms. Facebook is able to make these surprising connections because so many users have handed over their phone and email contact books, allowing the company to draw lines between people like constellations in the stars.

harassment and to avoid the penalties of stigma. "We're not a protected class," she said. In other words, people can discriminate against adult film actors. They can refuse to hire them, rent them an apartment, or provide services to them.

When a recent landlord found out about Darling's adult film work, she refused to renew Darling's lease, leaving her scrambling for a new place to live. A couple years back, Airbnb, the house rental site, banned Darling from its platform during a time when it appeared to be purging anyone who worked in the sex industry. Darling said she had never even had sex in an Airbnb. But this is the world facial recognition will usher in even more rapidly, one in which people are prejudged based on choices they've made in the past, not their behavior in the present.

DAVID SAID HE had no interest in outing the women or causing any problems for them. He considered himself a digital Peeping Tom, not interfering in any way, just looking through a window into a porn star's real life. He would screenshot the results of his searches and store them on an encrypted drive, because he never wanted anyone else to find them.

It was a solitary game for him, an invasive fetish that appealed in part because of the challenge it presented. Once he came across a woman he liked in a "casting couch" video, a genre of porn in which a prospective actress is interviewed and then "auditions" for the job by having sex with her interviewer on said couch. He ran the young woman's face through PimEyes. One of the hits was a photo of her on a high school website. Her name wasn't included, but David kept searching and discovered that the school had a Flickr account with thousands of photos. He found "spring formal" photos from the year he thought she might have graduated and started scrolling through them, and within twenty minutes, he found a photo of her that included her name. He was a technologist, but he felt he could change careers to become a private investigator.

Once David knew an actress's identity, he tended to lose interest in her. Eventually, he tired of unmasking professionals and moved on to what he considered truly shameful. "I went through my entire Facebook list of friends," he said. Not the entire list. Just the women.

Over fifteen years on Facebook, he had befriended hundreds of women. The first person he got a hit for was a near stranger he had met one time at a club while on vacation. They had become Facebook friends and then never interacted again. "It turned out she shot porn at some point in her life," he said. "She's a brunette now, but in the porn, she was blond."

Then he found more: A friend had posted nude photos to a Reddit community called Gone Wild, a place intended to anonymously collect compliments on one's body. There were topless photos of an acquaintance who had participated in the World Naked Bike Ride. A woman who had applied for a room he had rented out once had naked selfies on a revenge porn website. The women's names weren't attached to the photos. They had been safely obscure until a search tool came along that organized the internet by face.

It can be extremely difficult to remove naked photos of yourself from the internet. Search engines such as Google have free request forms to excise them from a name search, but what about a face search? That, naturally, was a service PimEyes provided—for a price. The PimEyes "PROtect plan" started at around $80 per month. It was advertised as a way to find photos you didn't know about, with "dedicated support" to help get them taken down from the sites where they appeared, but one woman trying to get regrettable photos removed from the service called it professionalized sextortion.

Originally created in Poland by a couple of "hacker" types, PimEyes was purchased in 2021 for an undisclosed amount by a professor of security studies based in Tbilisi, Georgia. The professor told me that he believed facial recognition technology, now that it exists and is not going away, should be accessible to everyone. A ban on the technology would be as effective, he said, as the United States prohibition on alcohol had been in the 1920s. Those who paid attention to a box you had to click before performing a search would see that you are only supposed to search for your own face. Looking up other people without their consent, the professor said, was a violation of European privacy laws. Yet the site had no technical controls in place to ensure a person could only upload their own photo for a search.

Too many people currently on the internet do not realize what is possible. People on OnlyFans, Ashley Madison, Seeking, and other websites that cultivate anonymity are hiding their names but exposing their faces, not realizing the risk in doing so. David wondered if he should tell his friends, anonymously, that these photos were out there, and findable due to new technology, but he worried that they would be creeped out and it would do more harm than good.

He had never uploaded his own face to PimEyes, as was the service's supposed purpose, because he did not want to know what photos it would turn up. "Ignorance is bliss," he said.

CHAPTER 21

CODE RED (OR, FLOYD ABRAMS V. THE ACLU)

When Nathan Wessler and Vera Eidelman found out about Clearview AI in early 2020, they immediately began thinking about what they could do to stop the company in its tracks.

Wessler and Eidelman are lawyers in the New York headquarters of the American Civil Liberties Union, the longtime foe of facial recognition technology. From almost the first moment the technology had shown a flicker of viability, the ACLU had been there, ready to fight, from the Snooper Bowl to Illinois to the campaign that got cities to ban police use of it. During the decades that facial recognition technology had been simmering in a pot, waiting to boil, the ACLU had been watching it. And now the pot was finally blistering hot, its contents bubbling over the edges.

Facial recognition was part of a larger agenda. The ACLU's Project on Speech, Privacy, and Technology had recently published a report with a spine-tingling title: "The Dawn of Robot Surveillance: AI, Video Analytics, and Privacy." It described the vast network of cameras around the world designed for constant capture and the photos and video accumulating online at breathtaking volume. That living record had been largely untapped, too much visual imagery for humans to review manually, but advances in artificial intelligence were changing that. It was becoming possible for algorithms to assess what was happening in real time in video footage and correlate all the streams of data coming in. The slumbering surveillance infrastructure was waking up. Clearview was just the first Godzilla-like creature to emerge from this collision of massive ongoing data collection with the supercharged analytical power of neural nets. There were more privacy monsters to come.

"Something that people don't generally think about a lot, but which is actually so core to our ability to just function in a free society," Wessler said, "is to be able to go about our lives, whether it's mundane or really embarrassing, sensitive or private things, and not expect we'll be identified instantaneously by a total stranger, whether that's police or some billionaire trying out a fun toy or anyone in between."

Wessler, thirty-seven, had been in college when 9/11 happened and had been alarmed by the way law enforcement targeted the Muslim community afterward. When he first joined the ACLU, he worked on national security cases; he represented a detainee at Guantánamo Bay and sued the U.S. government after it killed three American citizens, including a sixteen-year-old, in drone strikes in Yemen. More recently, he had switched to the issue of surveillance; he had helped convince the Supreme Court that people's location history, as gleaned from phone carriers, deserved legal protection; and he was part of the ACLU legal team representing Robert Williams in Detroit.

Wessler and Eidelman were troubled by what Clearview had built and how it might allow the state to track anyone whose face was captured by a camera, including antigovernment protesters. Eidelman, thirty-two, had been born in Russia and moved to the United States when she was a toddler. After working for Google briefly after college, she'd gone to Yale Law School, where she studied free speech and the First Amendment. She had recently helped block a law in South Dakota that would have chilled activists' right to protest a new oil pipeline. She worried that the government might utilize facial recognition as a tool of intimidation.

Though police use of mug shots and driver's license photos had long concerned the ACLU, there had at least been some rules and democratic control over the use of those databases. Clearview was a whole different beast. "Here was the Silicon Valley startup approach to this technology," Wessler said. "'Let's build a thing that no one else has wanted to build yet because it seems troubling and let's throw it out there and see where it sticks and where we can make money.'"

A conversation about Clearview started within the ACLU as soon as the company's existence became public. Typically, the organization sued the government to protect citizens' rights, but Clearview was part of a trend the organization was seeing: private vendors selling law enforcement agencies the fruit of surveillance methods that would likely be unconstitutional if deployed by the government itself. There was a company called Persistent Surveillance Systems that flew planes over cities to take detailed aerial photos so police could reconstruct events from a bird's-eye view. Vigilant Solutions, a vendor that took photos of license plates all over the country, sold the ability to quickly locate a car. Data brokers who siphoned location information from smartphone apps were selling a history of people's movements to federal officials. Police didn't need a warrant; they just needed to pay for the intel. Centuries-old rights protecting American citizens from government abuses were growing increasingly irrelevant as private companies took over the work of surveilling the country's inhabitants.

"The question was 'Can we bring a case and what would it look like?'" Eidelman said. It was more complicated than it appeared on its surface: Clearview represented a clash between free speech and privacy. The company had used freely available information—public photos, the use of which was arguably constitutionally protected—to build an unprecedented surveillance tool that potentially violated constitutional protections against unreasonable searches. Should the ACLU sue Clearview AI?

While the organization debated what it should do, private citizens sued Clearview. The first class action lawsuit was filed in federal court in Illinois at the end of January 2020, four days after the company's existence hit the front page of *The New York Times*. Clearview was accused of violating Illinois citizens' constitutional rights, as well as that all-important state law the ACLU had helped to create: the Biometric Information Privacy Act, or BIPA, which had brought the best-resourced technology companies in the land to their knees and cost Facebook $650 million.

"Almost a century ago, Justice Brandeis recognized that the 'greatest dangers to liberty lurk in insidious encroachment by men of zeal, well-meaning but without understanding,'" the class action complaint began, quoting the memorable Supreme Court justice who had sketched out the need for a "right to privacy" in 1890. The complaint said that Clearview "epitomizes the insidious encroachment on an individual's liberty that Justice Brandeis warned about" but that its creators "acted out of pure greed."

There were 12 million people living in Illinois. Even if Clearview had only a tenth of them in its database, that would cost $6 billion at $5,000 per faceprint nonconsensually taken. A loss would likely mean bankruptcy and the end of the company.

More class action lawsuits were filed in other states, accusing Clearview of invasion of privacy. In March 2020, Vermont's attorney general sued the company for violating consumer protection laws, calling Clearview's web scraping and face mapping "unscrupulous" and "unethical." And then, in May 2020, the ACLU decided to come off the sidelines.

Wessler and Eidelman, alongside local counsel and the law firm Edelson, which had brought the BIPA suit against Facebook, sued Clearview in Illinois state court, calling the company's technology a "nightmare scenario." It was a lawsuit on behalf of domestic violence victims, sex workers, and undocumented immigrants, who all had a greater sensitivity than most to the dangers of having their faces tracked. The ACLU wasn't seeking monetary damages. Its main objective was to vindicate BIPA, in hopes it would inspire other states to pass their own versions of the law to protect their constituents from the likes of Clearview.

Clearview was facing a legal onslaught. Class action lawsuits were to be expected, but the optics of being sued by America's leading civil rights organization cast an especially negative light on the company. Seeing the seriousness of what it was up against, Clearview decided it needed a heavy hitter of its own.

IN JULY 2020, Floyd Abrams got a call out of the blue from Richard Schwartz. Abrams, eighty-four, was senior counsel at Cahill Gordon & Reindel; he'd once been a partner at the firm but, given his age, now had the title often given to long-in-the-tooth legal legends who were no longer required to work long hours at the office. Not that he was at the office; due to the pandemic, he was working from home when he got Schwartz's call.

Abrams hadn't heard of Clearview, so Schwartz explained the company's product in broad strokes and the legal trouble it was facing. The company had already stopped selling its app in Illinois and Vermont and had attempted to keep Illinoisans from coming up in searches, a near-impossible feat.*

Clearview had a robust team of lawyers, but the company needed an expert on the First Amendment to argue that freedom of speech gave it the right to organize, analyze, and present information that was already public. If so, that would mean that any law interfering with Clearview's ability to do so, such as BIPA, was unconstitutional.

The argument wasn't necessarily intuitive. Scraping photos and resurfacing them was a form of speech? But it wasn't actually that far-fetched. Judges had drawn expansive circles around the types of human and corporate expression that were protected by the First Amendment, including computer code, political spending, video games, and web search results. "I found it really interesting," Abrams said. "Here we have twenty-first-century judges addressing twenty-first-century technology to see if they're consistent with an eighteenth-century document."

Schwartz asked Floyd Abrams if he would be interested enough to join Clearview's defense team and sent him a collection of the complaints that had been filed. "It took me a few days to get through the words 'facial biometric algorithms,'" admitted Abrams. He decided to take the case even though his long career had not involved much work in the area of novel technologies.

Abrams had first become famous in the 1970s, when he represented *The New York Times* in a legal showdown with the Richard Nixon White House over the newspaper's right to publish the Pentagon

*Clearview said that it had identified all scraped photos with location metadata indicating that they had been snapped in the state, as well as all photos with captions that mentioned Illinois, Chicago, or other cities in the land of Lincoln, and filtered them from results.

Papers—classified documents that exposed government lies about the Vietnam War.* The high-profile case went to the Supreme Court, which sided with the press's right to publish. It helped establish Abrams, then in his early thirties, as a First Amendment expert.

Since then, Abrams had argued before the Supreme Court more than a dozen times. He had written three books on the First Amendment. Yale Law School had dedicated an institute to him, and Vera Eidelman had participated in a clinic there that was formative in her decision to go into the same field.

Abrams was a legal lion, if an aged one, and Clearview AI added him to its circus. In the main ring: the legendary First Amendment lawyer against the world's premier First Amendment organization.

The ACLU had been formed to fight for the right to free speech, no matter how offensive, and had been protecting it for more than a hundred years, going to the Supreme Court even more times than Floyd Abrams. In the early days of the internet, the ACLU demolished a law intended to protect children that would have made it a crime to post sexually explicit messages and dirty pictures online. That was what made the Clearview case so tricky for the organization. If Hoan Ton-That had created a facial recognition app for use *on* police officers to hold them accountable and the government didn't like it, the ACLU might have defended his First Amendment right to create and distribute such an app.

A scenario like that arose in Oregon right around the same time. Portland, like many other American cities that summer, was the site of frequent protests following the police killing of a Black man named George Floyd. Police patrolling protests that were about *them* created volatile situations that sometimes turned violent. Some officers began hiding their badge numbers to make it harder to identify them, ostensibly for their own safety.

After being tear-gassed at a protest, a self-taught coder named Christopher Howell resolved to create a face recognition database to

*The *Times* had gotten highly classified documents from a whistleblower named Daniel Ellsberg revealing that the U.S. government had systematically lied to the American public about the true nature and toll of the Vietnam War.

identify these police officers, using a Google tool called TensorFlow.* But when the Portland City Council passed a law soon afterward prohibiting police, corporations, and private entities from using facial recognition technology to identify unknown people, Howell got worried that what he was doing was illegal. He attended a council meeting to ask whether his project violated the city ban. Portland's mayor told Howell that his project was "a little creepy," but a lawyer for the city clarified that the new rules did not apply to individuals.

If it had gone the other way and Howell had asked for help from the ACLU of Oregon, the group would have had a difficult decision to make.

VERA EIDELMAN HAD seen Floyd Abrams speak at Yale when she was a law school student. He was an icon, and she was around the age he had been when he took on the Pentagon Papers case. But his involvement did not come as a complete surprise.

Though Abrams had gotten his First Amendment bona fides working on behalf of journalists, he had spent much of his career since defending the speech rights of corporations. After the 2008 financial meltdown, he had argued that the solid gold AAA ratings that Standard & Poor's had given to junk debt were simply the company's "opinion" and not something it could be held liable for. In 2009, on behalf of a big tobacco company, he had challenged a law requiring "shocking color graphics" as warning labels on cigarette packaging, saying it violated the cigarette makers' expressive rights. He was also on the winning side in the 2010 Citizens United case arguing that capping corporations' political spending violated their free speech.†

Eidelman didn't feel intimidated, even though she had a nuanced argument to make. The ACLU didn't object to Clearview's scraping of

*One of the sources for officers' photos, beyond Facebook, was OpenOversight, a crowdsourced database for looking up law enforcement officers by name or badge number. It had been created by a nonprofit in Chicago co-founded by Freddy Martinez; the nonprofit endorsed Howell's use of the photos. The same people opposed to facial recognition *by* the police approved of its use *on* them, to hold them accountable.

† The ACLU had been on the same side as Abrams in the Citizens United case.

public photos. It was generally in favor of people being able to collect public information from the internet. What the ACLU objected to was Clearview's capture of a biometric identifier from those photos without consent. If the First Amendment became a means for corporate entities to justify any analysis of individuals, no matter how invasive or intimidating, people's privacy would be eviscerated.

IN SEPTEMBER 2020, Hoan Ton-That and Floyd Abrams met to discuss legal strategy and to do a demo of the app. They allowed me to observe, as color for a *New York Times Magazine* article I was working on, which came as a surprise to me. People involved in high-stakes lawsuits generally don't let journalists sit in on their legal strategy sessions. Their doing so can imperil attorney-client privilege.

Clearview still had no real office at which to meet, though it was otherwise becoming a more professional operation. It had hired high-powered attorneys from established corporate firms, not just Abrams but also Lee Wolosky, a former national security advisor, and was recruiting employees with experience in government contracting. Clearview's few dozen full-time employees were still working from wherever they wanted, and although a government contractor without an actual headquarters was oddball, it was less so in the pandemic; lockdowns and social distancing rules had emptied most corporate offices.

Instead, the meeting took place in Abrams's posh Fifth Avenue apartment in Manhattan, its foyer adorned with photos of Abrams alongside Barack Obama, Bill Clinton, and George W. Bush. Abrams had a home office there, but it wasn't huge, making it hard for us to maintain the recommended six-foot distance from one another in the time of Covid-19. Clearview's spokeswoman, Lisa Linden, arrived early in a floral mask, and Hoan Ton-That arrived a few minutes late, wearing a paisley jacket over an all-white ensemble, a red bandanna functioning as his mask. Ton-That set his gray laptop bag on the floor and reclined in a chair that seemed too small for his lanky body.

In the middle of the room were Abrams's desk and computer, a Ring light hovering above it like a hollow moon. Abrams was working on a motion to dismiss the ACLU lawsuit and asked Ton-That some questions about how exactly to describe the company's algorithm.

"Does one say the app makes judgments?" Abrams asked. It just provides information, Ton-That said, casually absolving his company of any mistakes an officer using the app might make.

Once the legal discussion wrapped up, Ton-That did a demo of the Clearview app, ostensibly for Abrams's benefit. But Abrams had already seen how the app worked, so the scene seemed staged, instead, for me. Ton-That moved to Abrams's computer chair to log in to Clearview AI. Ton-That googled Floyd Abrams's name, grabbed one of the photos that came up, and uploaded it to Clearview. Usually results appear instantly, but there was a delay, some kind of technological hiccup. Ton-That laughed nervously. "Maybe this is less dangerous than people think," Abrams quipped.

Ton-That decided to try again with a different photo. He googled Abrams's son, Dan Abrams, a legal correspondent at ABC News, and used a screenshot of him instead. This time, the app performed beautifully: The screen filled with a grid of photos of the younger Abrams from around the web, the source of each identified in tiny type under the photo. Ton-That clicked on one of the photos, where Dan Abrams was standing with a woman, then clicked on the woman, which brought up numerous photos of her as well.

They talked about how this could theoretically be done manually, that a company could hire thousands of people and task them with going through photos to find all the ones of a particular person.

"Nobody would say the state could come and limit that," Abrams said, remarking that Clearview was simply making these matches "in a far speedier and more accurate way than hiring five thousand people" and that it shouldn't be punished for using a computer to do so more efficiently. Ton-That said that he tried not to think about losing and that he needed to "just focus on growing the business to overcome the legal costs." He seemingly recognized it as the nature of the beast: Being a radical new startup entailed legal troubles. Uber had dozens of federal lawsuits filed against it, and hundreds at the state level. "They just kept collecting them," he said. "But they kept on raising the capital." Same thing with PayPal, the money transmission company cofounded by his investor Peter Thiel, he said; it got sued by twenty-eight state attorneys general. "What they ended up doing was jamming all those cases," he said, "and making them slow down as long as possible while the company grew."

Ton-That acted like it was all part of the plan. While the lawsuits progressed, Clearview would continue trying to raise money from investors. It would sell its product to new customers. And it would grow its database from 3 billion faces to more than 30 billion and counting.

MANY MONTHS LATER, in April 2021, Abrams got to make those same arguments to the judge in Illinois over Zoom, due to pandemic precautions. A lawyer for Clearview kicked off the hearing by arguing that the case should be dismissed because a law in Illinois shouldn't apply to a company based in Delaware that wasn't doing business in the state (anymore).*

Then it was Floyd Abrams's turn. He called it an "indisputable proposition that photographs on the Internet by their nature are not only public in nature, but that claims of loss of privacy by people who put them on the Internet are routinely rejected." If Clearview were prohibited from using technology to analyze public photos, Clearview could still proceed with its business but only by hiring thousands and thousands of humans to sort through photos, which he said would be perfectly legal but "not practical" and, with a billion or more photos, "impossible."

The judge expressed skepticism. There was no issue with Clearview's use of the photographs. "What we're really talking about is the use of this faceprint that is made from the photograph," she said. Abrams continued to argue that Clearview had a constitutional right to analyze a photograph however it chose, including deriving a biometric identifier from a face within it.

Vera Eidelman seized on that when it was the ACLU's turn. Despite some problems with her internet connection—she kept dropping out of the Zoom session and eventually called in by phone instead—she powered through her presentation to the court.

"We're not arguing that Clearview can't collect photographs from

*In Illinois, 105 police departments and companies had conducted thousands of Clearview searches.

the internet, we're not arguing that Clearview can't match photographs or express an opinion about who appears to be in a photograph," she said. "All we're arguing is that Clearview cannot use nonconsensually captured faceprints in order to do that. It can't capture the faceprints of our clients, individuals who are survivors of sexual assault, sex workers, undocumented immigrants, people who are regularly seeking to exercise their First Amendment protected rights."

Protecting Illinoisans' faceprints helped protect *their* First Amendment rights, Eidelman argued. If Clearview were used to track faces at a protest, for example, people might no longer feel comfortable assembling, chilling *their* free speech. According to Eidelman, Clearview's argument was dangerous and would make it impossible for Americans to protect other sensitive biometric data about themselves. "There isn't really a difference between the argument that one could extract a biometric identifier from a photograph online and the argument that one could dust for fingerprints anywhere in public or extract DNA from a Kleenex or water bottle that's left out in public," she said. "That's not what the First Amendment stands for. The First Amendment doesn't require that we have access to all information that exists."

The judge issued her ruling a few months later. She said that the First Amendment wasn't a get-out-of-jail-free card. A state could still outlaw certain kinds of data collection, including the collection of people's faceprints, whose use could cause harm to them. "The fact that something has been made public does not mean anyone can do with it as they please," she wrote. "The photos may be public, but making faceprints from them may be regulated."

She scoffed at Clearview's claim that it couldn't possibly get permission from every Illinoisan whose face it wanted to make searchable in its app. She said that the company should figure out a way and not expect to be exempt simply because it had "forged ahead and blindly created billions of faceprints without regard to the legality of that process in all states."

Vera Eidelman's arguments had been more convincing than those of Floyd Abrams. The judge said that the ACLU deserved its day in court against Clearview. Clearview's attempt to play the First Amendment card in its other lawsuits was similarly shot down. Clearview wasn't getting out of litigation that easily.

But the ACLU was evidently not keen to go to trial either. In May 2022, Clearview and the ACLU agreed to settle their lawsuit. Clearview didn't have to pay any money beyond the ACLU's legal fees; it just had to agree not to sell its face database to private individuals and companies in the United States. It could continue working with law enforcement, even in Illinois, where, it had argued, it was exempt from BIPA as a government contractor, but it agreed not to sell its product to Illinois police for five years. Clearview boasted that it was a win, as it allowed the company to continue doing what it already had been doing: selling facial recognition to the government.

The next month, Hoan Ton-That organized an appreciation event in a private room at Michael's, a restaurant for power players in midtown Manhattan. He dubbed it Lawyer Appreciation Day, inviting the company's legal teams from around the world to attend. "Please don't bill me," he wrote in the invitation. Clearview had new marketing and sales executives that it had hired away from Vigilant there, two in-house lawyers, and one of its investors. (It wasn't David Scalzo, who did not appreciate all these lawyers one bit.)

Lisa Linden had invited a photographer, and Ton-That jauntily posed for photos in a powder blue suit. I was also there, hoping to meet Richard Schwartz in person for the first time, but he was nowhere to be seen, seeming to want, as always, to stay in the shadows.

Only Clearview's American and Canadian lawyers showed up to sip the event's signature drink, "subpoena coladas" the shade of Clearview blue.

CHAPTER 22

THE FUTURE IS UNEVENLY DISTRIBUTED

When does a pandemic come to an end? For England, it was January 27, 2022. On that Thursday, many Covid-19 precautions were officially eliminated. Brits would no longer need to wear masks over their noses and mouths. They could throw caution and germs to the wind.

The following day around noon, as the city bustled with the seeming return of something like normalcy, London's Metropolitan Police drove a white Iveco van—the kind popular with #Vanlife enthusiasts—to a pedestrian-friendly shopping district called Oxford Circus. Mounted on the van's roof were two high-resolution cameras, and inside was computer equipment, including two gaming laptops, designed for fast visual processing.

The police parked near an entrance to the Oxford Circus Tube station, where subway travelers would be flowing in and out. Their spot was across the street from an abandoned Topshop. The clothing retailer had gone out of business during the pandemic, and while the building awaited a new occupant, its large storefront windows were covered over with bright green paper that said BE THE FUTURE over and over again in a hypnotic black-and-white swirl.

The officers set up a red sandwich board on the sidewalk about fifteen feet from the van. It read POLICE LIVE FACIAL RECOGNITION IN OPERATION in all-caps white letters and explained that officers were looking for people "wanted by the police or the courts."

The Metropolitan Police, or the Met for short, had compiled a watch list of more than nine thousand people that it had some reason to believe might walk by that particular spot that particular day. How the officers had created the watch list and who exactly was on it were known only by them, but they would later write in a memo defending their legal right to do so that live facial recognition technology "helps the Metropolitan Police Service (MPS) locate dangerous people who are wanted for serious criminal offences. It helps us keep Londoners safe."

It was the fourteenth time the Met had rolled out its mobile facial recognition unit since August 2016, when it had test-run real-time scanning of faces for the first time at a carnival in Notting Hill. That had been a troubling choice. The carnival was an Afro-Caribbean celebration, so it looked as if the police had purposefully selected an event heavily attended by Black people.

Oxford Circus was a more neutral location, almost as crowded as the street festival but with the shoppers and pedestrians in winter coats. The Met had a mandate to keep the public informed about its use of the controversial technology, in honor of the country's democratic principles. So just an hour before the deployment, at 11:00 A.M., the police tweeted that they would be using facial recognition technology that day "at key locations in Westminster." That, along with the red sign, was, to their minds, sufficient public notice.

For Silkie Carlo, those tweets were like a fire alarm going off. As soon as she saw one of these announcements from the Met about a facial recognition deployment, she would drop whatever she was doing. Carlo, thirty-two, ran Big Brother Watch, a small privacy-focused nonprofit in London. The daughter of a tour manager for Led Zeppelin, Carlo had long blond hair parted down the middle and a bluebird tattoo on her wrist. An elegant and photogenic civil liberties crusader, she was frequently invited onto news programs to make the case against excessive government surveillance, a sometimes frustrating exercise in the CCTV-friendly United Kingdom, where polls regularly found that the public favored security measures over privacy concerns.

When Carlo saw the Met's tweet, she was on the phone with the United Kingdom's biometrics and surveillance camera commissioner, a regulator tasked with making sure authorities used these technologies appropriately. The subject under discussion was the surveillance vendor, Hikvision, a Chinese camera and video analytics company. Its products, at the perfect intersection of high quality and low cost, were extremely popular. But they had been used to surveil Uighurs and other Muslim minorities in China. The company was on a U.S. government sanctions list. Carlo argued that by using the cameras, the United Kingdom was condoning and financing human rights abuses abroad.

She stopped mid-discussion, told the commissioner she had to go, and alerted her small band of co-workers—there were four of them and an office dog named Alfie—that it was time once again to gather their protest signs and literature and find out exactly where in Westminster the police van was scanning faces so that they could go make a fuss about it.

Oxford Circus, London's busiest pedestrian area, was the first place they checked. They arrived around 2:00 P.M. and saw, as they usually did, dozens of uniformed and plainclothes officers standing around, all staring at iPads, waiting for a match to come up. The officers were supposed to wait for their colleagues in the van to vet a match, because the facial recognition software, from the Japanese tech conglomerate NEC, wasn't foolproof and didn't claim to be. The algorithm would say, for example, that a passerby's face had an 85 percent similarity score to the face of someone on the watch list, but the officers in the van would rule the match out if, for example, the passerby was a woman and the wanted person was a man.

But there was a risk in waiting too long: A person might fade into the crowd and be hard to find again. So the mass of officers outside the van were eager to accost anyone who lit up their screens—or anyone who tried too obviously to avoid the cameras.

Carlo stood outside the Oxford Circus tube station holding a sign with a big white arrow that read, THIS WAY TO BE BIOMETRICALLY SCANNED BY THE POLICE. She was a familiar—and annoying—presence at the facial recognition deployments. She handed out leaflets to people walking by and answered questions about what was happening and why it troubled her. London was already one of the most highly surveilled cities in the world, having rolled out CCTV cameras extensively in the late 1990s and early 2000s in reaction to bombings and terrorist activity in the city. The urban landscape was dotted with cameras and signs warning people that the cameras were watching them. Carlo's big fear was that the trials would be deemed successful and that London, and perhaps all of England, might decide that tracking the faces of the more than 50 million people who lived there, as well as any visitors, was a peace of mind the country could not live without.

She hoped that by being at Oxford Circus on that cold January day, she was making that Orwellian future less likely. "The facts we turn up cause friction," she said. "And our presence there means they're getting fewer matches." She wanted to witness any mistakes by the police or the technology so that she could publicize them and sway the debate. She got what she'd come for.

At one point, a group of officers surrounded a Black teenager and questioned him for twenty minutes. The facial recognition had flagged him as a potential match for a wanted person. The alarmed teen was too young to have a government ID on him to establish his identity. Eventually the officers fingerprinted him and only then confirmed he wasn't the man on the wanted list that the facial recognition system had suggested he might be.

Carlo saw another person who was stopped after being correctly identified by the facial recognition system. But he shouldn't have been stopped. He was out on bail, awaiting a trial, and not "wanted." That is a frequent problem with watch lists: Once you are put on one, it can be hard to get off.

According to the Met's official report, the faces of more than twelve thousand people were scanned that Friday afternoon. The facial recognition alert went off eleven times. The Black teenager's engagement became a statistic in the report: one confirmed false alert.

The Met said that the other ten had been "true alerts," that seven people were stopped, and four people were arrested. They considered it a success in their "ongoing work to tackle serious and violent crime." They had chosen that random block on that random day with a watch list of thousands of people, and in a matter of hours had managed to find and identify a handful of them.

They put out a press release about the dangerous men who were arrested: a thirty-two-year-old who was to be extradited on charges of assault and drugs, a man wanted for making death threats, a thirty-one-year-old who had a warrant out for drug offenses, and a forty-year-old who had a warrant out in relation to a traffic offense.

Yes, the London police were arguing that facial recognition was needed to find rapists and murderers, but here it was being used instead to track down a traffic offender. Had the man sped or driven under the influence, or did he just have an unseemly number of unpaid parking tickets? The Met would not say.

In May 2022, I went to London to get a glimpse of the city at that crucial moment. Would Silkie Carlo convince her country that facial recognition technology was too chilling or too flawed to use regularly? Or would the Met be granted license to move its live facial recognition technology from mobile vans to the city's ubiquitous cameras?

Almost as soon as I landed at Heathrow Airport, I had to submit to a version of the technology—the type generally regarded as the most innocuous, a one-to-one comparison to verify that I was, in fact, me. After I got off the plane on a Sunday around 10:00 P.M., I waited in line at an electronic passport gate. When it was my turn, I placed my passport on a red-lighted digital bed and looked into a camera. Within seconds, an automated system scanned my face to match it to a biometric identifier embedded in a chip in my passport. The gates opened, and I easily entered the United Kingdom, potentially leaving my face behind. (The Home Office refused to tell me whether they keep images of travelers who pass through the gates.)

The next morning, I met up with Pete Fussey, a professor of sociology at the University of Essex who focuses on surveillance. He had convinced the Met to let him embed with its mobile facial recognition unit for six deployments over 2018 and 2019 so that he could give them a candid human rights assessment of their policies and procedures.

Fussey and I met for coffee at one of the locations where he had watched the watchers, outside a mall in a residential neighborhood called Stratford. There is a pedestrian bridge leading to the entrance of the mall that had created a perfect funnel of faces to scan. The bridge connects to a diverse neighborhood that the police described as having high rates of street crime, elevating the chance that a wanted person might walk by, according to the Met's reasoning.

Fussey had been given unparalleled access to the police during these deployments. He had sat beside officers at the gaming laptops, reviewing the matches coming in. He watched as they decided whom to stop, question, or arrest. I was not granted the same access. The Met and the South Wales Police, which was also experimenting with live facial recognition technology, declined to meet with me or demonstrate their technology. A spokesperson for the Met pointed to the "large amount of information" available on its website, and the South Wales Police said it was in a "pilot and evaluation stage" during which it was not engaging with the media.

It's possible that Fussey was partially to blame. The hundred-plus-page report he had written, together with a human rights law professor, about the Met's facial recognition trials had raised myriad concerns—about how the police came up with their watch lists, how they treated possible matches, how they treated people who tried to avoid the cameras, and whether it was all legal. To comply with human rights law, a new invasive surveillance technology needed to be "legal, legitimate, and proportionate," and the report concluded that scanning thousands of people's faces was not worth the one person that might be caught.

"I think live facial recognition makes us less safe," Fussey said. Watch lists were destined to swell in numbers. False positives were inevitable. Officers would be focused on the watch for wanted people rather than putting resources into more pressing issues. He believed that having tens of thousands of cameras constantly on the hunt for low-level offenders would divert officers from more important and fundamental work.

Police weren't the only ones using facial recognition in England. A high-end casino called Les Ambassadeurs, housed in an ornate four-story former royal hunting lodge near Buckingham Palace, had four hundred cameras, ten of them equipped with facial recognition from an Israeli company called Oosto. The face-scanning system had been installed only a few years earlier after the casino's head of security deemed the technology accurate enough to be worthwhile, around the same time police began testing it. The cameras were used to identify the membership-only casino's top-tier guests as soon as they walked near the premises to provide them with "white-glove service."

The system was also used to keep out unwanted visitors. The casino had a watch list of nearly fourteen hundred people who were to be turned away if they tried to enter. They included a handful of guests who had caused problems at Les Ambassadeurs in the past; more than three hundred people accused of cheating, abuse, or violence at casinos in the country; and approximately a thousand self-identified gambling addicts who had asked to be barred from places where they could lose their money.

"The primary use case is customer service," the casino security chief said. "Not all technology needs to be bad or detrimental. It's how you use it."

In the last few years, facial recognition technology has also been deployed in grocery markets, gas stations, and corner stores across the United Kingdom as a means to thwart potential thieves, thanks to a security company called Facewatch. The London-based company allows retailers in its network to scan the faces of their customers and get alerts when the system spots people who have been flagged by other merchants for shoplifting or being abusive to employees, like a neighborhood watch for the commercial sector. Facewatch claims to reduce theft by 35 percent.

When I met Silkie Carlo in May 2022, she had recently shopped at a Southern Co-op grocery store known to be using Facewatch. She had gone there specifically so that she could file a complaint with the U.K. privacy regulator about the system's use of her biometric information. She was a one-woman dam trying to hold back the technology in her country.

IN THE EARLIEST days of planning the company that would become Clearview AI, Ton-That got wind of a facial recognition app in Moscow that was causing headlines. He had sent an article about it to Richard Schwartz and Charles Johnson, commenting, "This will be the future."

With FindFace, you could take a photo of anyone, and the app would suggest public profiles of similar-looking people from VKontakte—VK for short—Russia's version of Facebook. The app used a facial recognition algorithm created by NtechLab, a little-known Russian company with just four employees. FindFace was essentially what Facebook had created but had not dared release.

FindFace went viral in March 2016 when an amateur photographer posted on VK that he'd used FindFace to track down two women he'd taken a photo of while out on a walk six years earlier. He said that he'd contacted the women and sent them the photo and that they had been delighted. In a VK post, which read suspiciously like advertising, he described FindFace as a "fucking cool algorithm," like "Shazam for people," and included a link to the service.

A few weeks later, another photographer named Egor Tsvetkov used FindFace on people who had sat across from him on the subway: a man with a shaved head listening to something on earbuds; a curly-haired brunette in a fur-lined coat staring at her phone; a blond woman, her hair plastered back, in heavy makeup, her eyes cast downward. He posted their subway portraits on his website alongside the photos that FindFace had unearthed.

The contrast was startling: The commuters were closed off and distant, in their own heads, focused on getting from Point A to Point B. The social media versions of them, however, stared into the camera, eager to connect, offering tantalizing details about their lives. Now the man with a shaved head wore a beanie and held the paw of a leashed, muzzled bear. The curly-haired brunette was revealed to have startling blue eyes, a small smile on her face over a candlelit dessert. The severe blonde turned welcoming, wearing a low-cut top and with her hair in a loose ponytail while holding a bottle of champagne. Tsvetkov called the project "Your Face Is Big Data," and it, too, went viral.

Three days later, FindFace took a darker turn. Users of a Russian message board called Dvach, "2chan" in English, started a thread called "Looking for Sluts Who Acted in Porn or Worked the Streets." They used the app to identify porn actresses as well as sex workers whose photos appeared on a site called Intimcity, and then, after finding VK profiles, messaged their husbands, partners, families, and friends about their alleged sex-based work.

The technology continued to be used by the Russian public, unchecked—sometimes for good, to identify two arsonists who were caught via CCTV setting fire to an apartment building—but often more controversially. A year after Dvach used it to out sex workers, a pro-government vigilante group called Je Suis Maidan used it to identify anticorruption protesters. Based on images in news coverage, Je Suis Maidan posted on its website the names and photos of about seventy-five people out of thousands who had attended a protest. (At least one person said that Je Suis Maidan got it wrong and that he hadn't been there for the protest.) Within a year of being made publicly available, a facial recognition app for identifying strangers was being used for every worst-case scenario imaginable, just as privacy activists had long warned.

Eventually, in the fall of 2018, NtechLab shut FindFace down, deciding to sell its algorithm only to the government and corporations. When Moscow began rolling out facial recognition across approximately two hundred thousand cameras as part of a "safe city" initiative, NtechLab was revealed to have been paid $3.2 million to help law enforcement more easily find missing children and arrest violent criminals.

Anna Kuznetsova had been in high school when FindFace first came out. She had an unpleasant experience with it on the Metro one day when some young men took her photo and then taunted her, showing her what they had found, including photos from when she was ten years old. In the summer of 2020, Kuznetsova, now twenty years old and in college, saw ads on the social messaging app Telegram for the ability to search for a face via Moscow's citywide facial recognition system. For 16,000 rubles, the equivalent of $200, she could send photos of someone and get a report on anywhere that person had been spotted by cameras. She decided to give it a try. When the system had been rolled out, the public was told that only government agencies would have access to the faces captured on streets, in public buildings, and on subway cars. But somehow, a black market for the data had sprung up.

Kuznetsova was not a jealous girlfriend tracking a significant other's whereabouts but rather a volunteer for a civil liberties group called Roskomsvoboda. They were investigating the abuse of Moscow's face surveillance system. Kuznetsova told the anonymous seller that she was looking for a person she used to know but sent instead a handful of photos of herself. She paid 16,000 rubles in the cryptocurrency Bitcoin and then waited. Two weeks later, her phone buzzed with a thirty-five-page report, showing each time the system thought it had seen her face on a surveillance camera, along with the location of each camera, the date and time she had been spotted, and the degree of confidence that it was her. The report contained more than three hundred sightings over the previous month; she did have a few doppelgängers in Moscow, but nearly a third of the sightings were of her. Some of the matches astounded her; the software had identified her while she was wearing sunglasses and a mask to protect against Covid. The system also accurately predicted where she lived and where she worked. She was horrified. It was a stalker's dream and a civil libertarian's nightmare.

Roskomsvoboda publicized what had happened and asked for a moratorium on the facial recognition system, arguing that it violated a privacy law in Russia that requires consent to use biometrics. Local authorities investigated, but there were no consequences for the facial recognition system writ large, only for the officers who were deemed to be involved in the illegal sale of the information. The officer who had released the data about Kuznetsova wasn't even supposed to have access to the system. He'd gotten a log-in name and password from someone else, showing just how lax the data controls were.

The system remained in place. There were errors. Roskomsvoboda heard from people who had criminal look-alikes who would be stopped frequently by police and have to show their passports to prove their innocence. But the public at large did not seem very concerned about the privacy intrusion. The Moscow police had a very high crime-solving rate, and many people appreciated the safety of the city. By 2021, the system was being used to identify antigovernment protesters, who could be arrested for the "crime" of participating in an unsanctioned public gathering.

Kuznetsova hoped that Russia could still find a "golden middle" with the technology, harnessing its benefits for solving serious crimes but not having it pervade day-to-day life completely. "Every society needs to decide for itself what takes priority," she said, "whether it's security or human rights."

CHINA HAS CHOSEN security. Paul Mozur first moved there in 2005. He was getting a degree in East Asia studies at Dartmouth and spent six months in Beijing to learn the language. He loved the chaotic energy of the place and resolved to return as soon as he could.

"I lived in a dorm with a bunch of mosquitoes and shared squat toilets, which was very affordable for a college student," he said. "Beer was cheaper than water at the time. It was glorious."

Just over a decade later, the city upgraded its bathroom tech considerably, flushing privacy away in the process. In 2017, toilet paper dispensers with facial recognition technology were installed in public bathrooms at the historic Temple of Heaven to thwart tissue paper thieves who had been squirreling rolls away in their bags. The machines dispensed two feet of toilet paper per face and would not give more to the same person until nine minutes had elapsed.

Mozur, a *New York Times* reporter, has covered China's rollicking tech industry since 2007 and has had a front-row seat to the nation's startling transformation into one of the most technologically advanced and tightly controlled countries in the world.

In the early 2000s, China was called the "world's factory" thanks to its cheap labor market. Among the many products being manufactured, assembled, and shipped out of the country were the component parts of computers, smartphones, e-readers, and gaming systems. China was viewed by many as the workhorse of the world economy—the hands, not the brains—but Mozur saw something different as he chronicled the rise of China's own, now-mammoth companies and apps, such as Xiaomi, WeChat, and TikTok.

Being at the center of the global technology industry catapulted China into the future ahead of everyone else. Mozur began seeing cameras installed everywhere. In Shanghai, he had to descend an escalator to get to his local grocery store, and there was a camera at the bottom aimed directly at his face "like a bazooka. It was not subtle," he said. "The cameras were like baroque sculptures, almost mocking surveillance. You'd have a camera hanging from a camera hanging from a camera on top of a camera. They were just everywhere, they were in your face, hanging low, literally at head height, looking into your eyes."

The United States and United Kingdom had installed surveillance cameras at the turn of the twenty-first century, putting them high up for a bird's-eye view of what people were doing. But that angle didn't capture faces very well, making it difficult to repurpose them for facial recognition technology. China, on the other hand, built its camera surveillance state just as facial recognition technology was becoming accurate enough to deploy widely, and it placed its cameras accordingly.

In major cities, authorities were able to create alert systems to locate people of interest in real time. And those people weren't just murderers and rapists. In the seaside town of Xiamen, the authorities flagged the faces of unlicensed tour guides so they could run them out of town. "Some of the use cases feel kind of silly, but in China, there's such a priority on order and the government being in control," Mozur said. "So charlatan tour guides are a nuisance."

Facial recognition cameras were used to deter petty offenses and "uncivilized behavior." In Suzhou, a lakeside city dubbed the "Venice of the East," the cameras were used to name and shame people wearing pajamas in public. The urban management department of the city of 6 million posted the leisure wearers' faces to social media, along with their names and, of course, the offending pajamas.

In Ningbo, the cameras captured people jaywalking and then sent them fines by text message, as well as projecting their faces on digital billboards to embarrass them. (A Chinese executive who had never been to the town was accused of jaywalking after the cameras captured her face in an ad on the side of a bus.)

Only VIPs could get onto a "red list" created by government authorities for people who should be invisible to the facial recognition systems. In China, being unseen is a privilege.

Mozur covered China's surveillance industry, going to sales conferences where vendors happily described what they could do, including "minority identification." In China, that meant mainly Muslims, whose population was concentrated in the country's western region of Xinjiang. The Uighur Muslims, in particular, had long advocated for independence from China. There had been riots in Xinjiang in 2009 and a series of violent attacks by Uighur terrorists, culminating in a Saturday-night stabbing of more than a hundred people at a subway station in March 2014. Since then, China had monitored Uighurs closely, blanketing streets and mosques with surveillance cameras and creating a vast network of detention camps to isolate and "reeducate" hundreds of thousands of people deemed to be "religious extremists." Their comings and goings were tracked relentlessly.

Facial recognition was just one tool in a vast and growing collection of surveillance technologies. The Chinese government had also started collecting DNA and tracking phones, layering multiple systems in an attempt to create an all-seeing eye on its population of 1.4 billion people. The government uses this eye for intimidation as well as monitoring, in a high-tech realization of the Panopticon envisioned by the eighteenth-century English philosopher Jeremy Bentham. The Panopticon was a prison in which inmates were always observable but didn't know whether the guards were actually watching them. Even the illusion of constant surveillance, Bentham theorized, would force them to behave.

All of this might seem like the stuff of horror novels, but for China, said Mozur, there was a utopian vision built into the surveillance, of a safer, more orderly society. "Tech became so linked culturally to the idea of China's progress economically that nobody was really scared of these things. There is just not the same kind of fear attached to the development of technology that we have in the West," he said. "It's just 'Holy crap. Look at what we can do now! Look how much richer we are! We're becoming like the United States! We're catching up!'"

But China was shooting far ahead, its camera and video analytics companies—SenseTime, Hikvision, Huawei, Megvii, Yitu, Dahua—becoming the largest and most advanced in the world thanks to deployment on a large scale in a densely populated country. But it wasn't one big, living, breathing network; surveillance was distributed regionally and controlled by local police.

In 2019, a woman in Shanghai complained online about the police installing facial recognition cameras in her apartment building's elevator. One of the things that bothered her was that the cameras had been subsidized by an advertising agency and were accompanied by a screen that played ads. The building's housing board had not had any say in the matter, and the woman was trying to organize a movement among residents to rip out the system.

Within an hour of the *Times* reaching out to the woman for an interview, local authorities contacted the newspaper's office saying that they knew reporters were interested in their "smart city" initiative. They had evidently intercepted the messages.

"We freaked out," Mozur said. "We realized we didn't have a lot of time to interview this woman. They are going to interfere with it."

To make sure he could not be tracked, Mozur turned off his phone and put it in a Faraday bag, a shielding mechanism that prevents signals being sent from or to a device inside. He hailed a cab, hopped out at a random location, paid in cash, and then hailed another cab and made his way to the woman's apartment to interview her.

He didn't take the same precautions the second time he went to see her. She was planning to go to the police station to talk to authorities about the cameras, and Mozur was accompanying her to observe. But as soon as they walked into the station, the police said they're weren't taking media requests. The police knew they were coming and exactly who they were.

This stunning level of authoritative scrutiny doesn't just impede the work of investigative journalists. The all-seeing eye of modern technologies has become a problem for America's spy agency as well. In 2021, the Central Intelligence Agency, which had funded that early and unwieldy form of face matching in the 1960s that depended on the efforts of teenagers, warned its operatives around the globe that informants were being captured and killed with alarming frequency due to the growing use of hacking tools, AI, biometric scans, and yes, facial recognition.

CHAPTER 23

A RICKETY SURVEILLANCE STATE

Soon after the 2020 U.S. presidential election was called for Joe Biden, rumors spread on social media that it would be possible to "stop the steal" and keep Donald Trump in office by disrupting a congressional ceremony in which lawmakers would certify the election results on January 6, 2021. That Wednesday, thousands of Trump's most fervent fans gathered for a protest outside the White House, where speakers, including prominent Republican lawmakers, claimed that the election had been illegitimate and urged the crowd to act.

"We will never give up, we will never concede. It doesn't happen. You don't concede when there's theft involved," Trump said from a podium flanked by American flags. "We fight like hell. If you don't fight like hell, you're not going to have a country anymore."

Then the president told the riled-up crowd to walk down Pennsylvania Avenue to the U.S. Capitol building to give the lawmakers assembled there "the kind of pride and boldness that they need to take back our country."

March the crowd did, and, upon reaching the Capitol, it overpowered police officers, broke windows, forced open doors, and surged inside the building. Lawmakers and their staff huddled in fear as they waited to be evacuated. Five rioters died during the attack, one shot by a police officer when she thrust herself through a shattered glass panel next to a barricaded door. Many officers were injured, one later dying from a stroke. After three harrowing hours, the National Guard arrived to clear the building. The mayor of Washington, D.C., announced a 6:00 P.M. curfew to keep people off the streets so that lawmakers could reconvene at the Capitol to certify the election in the wee hours of the night.

It was a shocking turn of events—an assault on democracy itself—and in order to identify those involved, some investigators turned to a tool hatched at the Republican National Convention more than four years earlier: Clearview AI.

In keeping with Trump's disdain for wearing masks, few rioters in the attack had worn them, despite being in the midst of a pandemic. Many of them thoroughly documented the unprecedented takeover of the federal legislative building on social media, leaving an abundant collection of videos and images that helped the FBI pick out the worst perpetrators. After they released photos of people wanted for questioning, Clearview saw a surge in searches, according to Ton-That, as local police departments around the country stepped in to help.*

Ton-That called the January 6 attack on the Capitol "tragic and appalling." The onetime Trump supporter said he was pleased to see his tool being deployed to mete out punishment to those who didn't want to see Trump leave the White House. He admitted that he had been "confused" before, and blamed the internet, but now described himself as apolitical. "In 2017, I was like 'I don't want anything to do with this. I just want to be building stuff,'" he said. "And now I never have spare time."

It was not very reassuring to think that Ton-That had pulled himself out of the authoritarian movement coalescing around Trump because he got too busy building an app that is, potentially, a powerful threat to civil liberties. Ton-That expressed hope that Clearview's assistance in prosecuting those who stormed the Capitol would make critics of the company come around. "You see a lot of detractors change their mind for a somewhat different use case," he said. "We're slowly winning people over."

At the very least, the work done around the attack on the Capitol helped win over financial backers. In July 2021, Clearview got

*The FBI did not officially have access to Clearview at that point, waiting until December 2021 to sign an $18,000 contract with the company.

a $30 million infusion from investors who valued the company at $130 million. It was continuing to ignore the punishing fines levied by European privacy regulators, and it still faced the prospect of having to pay Illinoisans $5,000 each, but the company kept chugging along.

Law enforcement agencies were aware of all the controversy around Clearview and facial recognition technology, and some investigators worried that they might lose access to a tool they increasingly considered indispensable. To prove that the technology could be used responsibly, Armando Aguilar, the assistant chief of the Miami Police Department, one of the local police forces that had assisted the FBI in its hunt for rioters, invited me to spend time in his Real Time Crime Center (RTCC) so I could observe how his detectives used Clearview and the other surveillance tools in their arsenal. It was illuminating, but not for the reasons he'd anticipated.

ARMANDO AGUILAR DOESN'T look like a character from *Miami Vice*. He doesn't have designer stubble, an all-day-at-the-beach leathery tan, or a white blazer. He has the pallor of an office worker; close-cropped dark hair, graying at the temples; and a general preference for the standard-issue black police uniform.

But the former homicide detective did have a starring role on TV for a while, on an early-2000s A&E series called *The First 48,* which tracked real-life murder investigations in the crucial first two days after the victim was found. The Miami police eventually pulled out of the production, because the documentary material it created could interfere with the slower, less showy business of prosecuting an alleged murderer, and because of the implication that the Florida beachside city was a dangerous place to live and visit. Aguilar moved on to what he calls "nerdy work" in the executive office of the police chief, then steadily climbed the ranks until he was put in charge of criminal investigations at the beginning of 2020.

It was around that time that he saw my *New York Times* article about Clearview AI. "We need to get this and set up a policy for using it responsibly," he told his colleagues. They told him that they had good news and bad news: They'd been using Clearview for months at that point, but they hadn't established any official rules. So he developed restrictions for facial recognition use: All officers would receive training, which would include a discussion of the tech's known biases, and its use would be limited to a set of core officers who had to record each search, and why it had been run, in a log that was regularly audited.

Miami detectives have since used Clearview AI to hunt down a serial booze thief who dressed as a county bus driver, a litterbug who Instagrammed himself throwing a bag of trash into the ocean, and a protester who threw rocks at the police, as well as murderers. "People are right to point out that this technology can be very dangerous," Aguilar said, comparing it to firearms and police vehicles. "Any of our tools can be dangerous without sound policy and training."

My three-day trip in March 2022 coincided with the Ultra Music Festival, an electronic music event that draws 165,000 fans to downtown Miami. Aguilar said the RTCC, the central repository for the department's real-time surveillance technologies, would be at its busiest because the festival paid overtime to have more officers on duty, both in the streets and digitally, to keep the event safe.

On the Friday I arrived at the police station, a four-story building in downtown Miami, Sergeant Alejandro Gutierrez greeted me at the front desk and showed me around. Parts of the building were musty with the kind of smells that are hard to dispel from Florida's heavy, humid air; one hallway smelled of urine while the waft of pot was heavy in an evidence intake room. There were SWAT vans parked in the back, but the tanklike BearCat was missing, said Sergeant Gutierrez, or Gooty, as his colleagues called him. It had been moved to make room for the police leadership's cars because the parking garage had recently been condemned after a piece of the ceiling fell off. Physical infrastructure seemed to be crumbling as the department's resources were pulled instead to the digital realm.

We took the elevator to the Real Time Crime Center, a dark, windowless box of a room lit only by the glow of computer monitors and screens that covered the walls. The center's logo was prominently displayed as we entered: a falcon with cameras for eyes in front of an

American flag with a thin blue line.* There were three detectives inside, two of them seated at a large shared workstation in the middle of the room and one at a desk in the corner, his workspace adorned with Spider-Man and Deadpool figurines. All of them wore guns in their holsters even as they spent the day staring at their computers, alternating between checking their email and clicking through feeds from the city's 583 surveillance cameras. Occasionally, officers on the ground radioed them when they needed their God's-eye view.

The huge screens covering every wall were mostly window dressing, Gooty admitted, more for visitors and police leaders who dropped in than for the RTCC detectives, who relied primarily on their computers.

One screen was displaying Snap Map, a service of the social video company Snap, available for anyone to access, that showed geographically tagged Snapchats recently posted by local users. The ones from Ultra that weekend were mostly frenetic shots of the crowd dancing under strobe lights, the Snap-using fans themselves becoming additional surveillance cameras. "This generation posts everything," one detective said. "It's great for police work."

One time, someone had posted a Snap of someone else hitting an officer with a skateboard. "We Clearviewed the guy and then passed the information along to the right unit," Gooty said.

On the other side of the room, tucked in a corner, was a flowing grid of license plates, beamed in from twenty different locations around the city. Some of the plates had a red box around them, indicating an alert of some kind on the vehicle. A RapidSOS screen had a list of phone numbers that had recently called 911 and from where. And there was a large TV, which was always tuned to Fox News.

The most interesting interactive screen in the room was devoted to ShotSpotter, a gunshot detection system with acoustic sensors placed in key locations in the city, for which the department paid $800,000 a year.

* The thin blue line is a symbol for police of what they do: holding the line so that society doesn't descend into chaos and mayhem. But "Blue Lives Matter" flags bearing the line have come to be seen by some as entrenchment in the face of criticism of police brutality.

It made a sound like a ray gun in a sci-fi movie each time a shot was detected and displayed a map of Miami that showed where alerts had gone off over the last twenty-four hours. Gooty tapped on a report of twelve rounds being fired in the wee hours of the morning in a majority-Black neighborhood called Overtown. The system let him play the captured audio. It sounded to me like fireworks, but ShotSpotter had labeled it "High capacity" and "Full auto," suggesting that it was from an assault rifle. Gooty said it sounded like fireworks to him as well but that shooters sometimes cover up gunshots with fireworks. Regardless, anytime ShotSpotter detects what it thinks is a gunshot, a police car is sent, sirens on, to check it out. "The system becomes our 911 caller," Gooty said.

Most of the center's other screens showed grids of surveillance camera footage, most of it from the streets around the music festival's many stages. The outside venue, and its enormous crowd, was just a half mile away.

By Friday evening, there were tens of thousands of people lined up to get into the Ultra festival for the opening set. A detective took control of a nearby camera and remotely panned it along a long, snaking line that had spilled into the street and surrounded an empty cop car. The detectives got a little nervous about that. "Let's keep the camera pointed this way in case something happens," someone said. When a person in line placed a Gatorade bottle on top of the car, everyone grumbled.

The center operated from 6:00 A.M. to 3:00 A.M. daily, and with Aguilar's blessing, I wound up spending twenty hours over three days in that room, the brain of a surveillance nervous system spread throughout the city. The police didn't do the slow, laborious work of in-depth investigations here. Rather, they sent messages out for others on the force, the body, to act on. They gave officers leads about what cameras had captured, whom a face belonged to, where a gunshot had gone off, who might have witnessed a crime.

It was both higher and lower tech than I had expected. There was a high-resolution, 4,000-pixel camera atop the forty-story InterContinental hotel next to the park where Ultra was happening. A detective could use it to zoom down to street level—to observe, undetected, what a festivalgoer was doing at an incredible level of detail. "It's like playing 'Where's Waldo?'" the detective joked.

But that camera was exceptional; the vast majority of Miami's cameras seemed to be far lower resolution and not as widely distributed as officers would have liked. Time and time again, when ShotSpotter went off, it would be in a blind spot, out of camera view. After the festival, each evening, the festivalgoers flooded out into the street and then seemed to disappear, flowing away from the camera-dense downtown to parts of the city that weren't as well covered.

An armed robbery happened on Saturday. Two guys were rushed by assailants with guns who took $3,000 in cash and a gold pendant. The robbers then got into a waiting car and sped off. It all happened right in front of a camera in a grocery store's parking lot, but the footage was so grainy that the detective couldn't read the car's license plate, and the faces so small and low-resolution that they would not have been worth running through Clearview. The footage was nearly useless, beyond capturing the time the robbery happened, basic bodily descriptions of the suspects, and which way the car may have gone.

The detective spent fifteen minutes or so reviewing other cameras in the area to see whether he could find the car zooming by but couldn't find it. That, too, was a surprise. Surveillance vendors were constantly saying that their AI could automatically search and scan video for a person, a car, or a face, but here was a detective having to look at car after car going through all the intersections around the robbery, trying to eyeball a match.

"Vendors always make these bold claims, but it rarely works as well as they say," Gooty said, which is why the department tends to insist on trials of a new technology, like the one it initially did with Clearview before signing up as a customer.

Aguilar stopped by to say hello, and shared a story about how camera footage had helped the department track down a real estate agent who had been attacking and murdering homeless men. They had tracked his drive after one of the attacks past various surveillance cameras, one of which had gotten a partial view of his license plate, which got them to the man's driver's license. They pulled his phone records, and his location history put him at the scene of the crime. "Thanks to all the technology, we got him at the bare minimum of being a serial killer," he said. "That guy would have killed a lot more people."

The roots of the Real Time Crime Center can be found in the terrorist attacks of September 11, 2001. The then newly formed Department of Homeland Security gave the city grants to buy cameras to set up downtown, where embassies and consulates were located (qualifying the surveillance as an "antiterrorism" measure). The Miami Police Department had been growing the network ever since, putting cameras on light poles and atop buildings, and in 2016 assembling the RTCC.

Miami was bringing about eighty new cameras online each year, but that will expand once the department's officers connect their body-worn cameras—initially put into place to hold police accountable—to the surveillance network. Their cameras will provide live feeds to HQ, making every officer a pair of potential eyes for the RTCC.

Given all the benefits that come from having such an extensive digital record of people's communications and movements, I was surprised by something Aguilar said that weekend. He admitted a concern about an overreliance on technology, and officers who thought they could solve a murder without leaving their desks. "You need to get out there and interview people," he said. "Technology has made us better, but we can't do it at the expense of traditional interrogative methods."

It reminded me of Pete Fussey saying in London that he thought live facial recognition technology would make us less safe. If officers rely on computers to collect the data, and then let a machine make decisions for them, and just act when an alert tells them to do so, they will miss things that only humans can see and do, and be trapped by AI's overreliance on what it has seen in the past to predict the future.

The only tech that proved critical that weekend at the RTCC was a surveillance camera pointed in the right direction at the right time. In my final hour there, an officer down on the streets of Miami radioed up from the scene of an accident. Both drivers were denying it was their fault. After the detective at the desk adorned with superheroes got the location, he found the nearest camera, a block away, and rolled back the tape. We watched as a dark SUV pulled up to a red light at an intersection, stopped as if it were a stop sign, and then kept going. A minivan coming fast, because it was on the road with the green light,

T-boned the SUV. The RTCC detective called the officer back to tell him that the SUV driver was responsible for the crash.

The surveillance state in Miami was incomplete, the cameras scattered and unreliable. It was not an all-seeing eye but a blinking, blurry one. To the extent that Miami represents what is happening in the greater United States, we have only the rickety scaffolding of the Panopticon; it is not yet fully constructed.

We still have time to decide whether or not we actually want to build it.

CHAPTER 24

FIGHTING BACK

Clearview kept looking for positive use cases of facial recognition technology to tout to the public and to sway detractors. When Russia invaded Ukraine, an act of aggression that most of the world decried, Clearview gave its facial recognition app for free to the Ukrainian government so officers could identify soldiers and spies. When it was reported that Ukrainians were using Clearview to find the social media profiles of dead Russian soldiers so they could send photos of their dead bodies to their loved ones, in order to convey the true toll of the war to the Russian public, *The Washington Post* called it "one of the most gruesome applications of the technology to date."

When a defense attorney reached out to Clearview asking to use the app to track down a Good Samaritan who was the only witness who could save his client from a vehicular homicide conviction, Clearview granted the attorney access. After it worked and his client was exonerated, Clearview said it would offer subscriptions to public defenders at a reduced rate if the technology could be similarly helpful in their cases. (This was not a violation of the agreement to only sell Clearview to the government, because many public defenders are paid with public funds.) But some in the public defense community were not impressed by the offering, calling it a PR ploy. Making a fundamentally unethical tool available to both sides, critics said, was not the solution.

The detractors refused to be won over. There is simply no use case for a technology like Clearview's that will sway someone like digital rights activist Evan Greer. "The hill I will die on is that face recognition is an inherently oppressive and harmful technology," Greer said. "We should think about face recognition the way we do about nuclear or biological weapons. It's a technology that has such a profound potential to erode the very fabric of human society that any potential benefits are outweighed by the inevitable harms."

Greer is part of a loose collective of activists, artists, academics, and technologists who are working desperately to contain facial recognition before it becomes as commonplace and unremarkable as other once novel technologies such as cars, email, and smartphones. Greer's organization, Fight for the Future, launched a "Ban Facial Recognition" campaign at the end of 2019 with more than forty other nonprofits, arguing that the technology—when employed to create surveillance dragnets—was simply too dangerous to exist. Dozens of civil liberties groups in Europe launched a similar campaign, called "Reclaim Your Face."

To draw public attention, Fight for the Future members dressed up in hazmat suits and strapped to their heads phones armed with a facial recognition app they had spun up using Amazon's Rekognition software. Then they wandered around Capitol Hill running the faces of everyone they encountered against the haphazard database of faces they had created, to try to identify them—just to make the point that there was no law barring them from doing so. They scanned the faces of more than thirteen thousand people and correctly identified a congressman from California. "Our message for Congress is simple: make what we did today illegal," the group wrote in a post afterward.

It isn't out of the realm of possibility that lawmakers at the federal level could rein in this technology. They have staved off a large-scale encroachment on privacy in public before by passing anti-wiretapping laws that protected people's conversations from being secretly recorded. That's why the millions of surveillance cameras around the country that watch Americans don't also listen to them. Elected officials around the country have also largely rejected another technology that would make police officers' jobs easier: speed cameras. We could set these cameras up on every road and automatically ticket each and every aggressive driver, but we have chosen not to, even if it results in more accidents on the roads.

It wasn't just activists pushing for regulations; even technology companies began asking for them. Faced with public unease about face tracking, the concerns of academics about racial disparities in the technology's deployment, and protests against police power following George Floyd's death, IBM, Amazon, and Microsoft, which all offered computer vision enterprise products, announced that they would not sell their face recognition algorithms to U.S. law enforcement agencies until the government came up with rules for ethical use.

Microsoft president Brad Smith said the pressure is building around facial recognition technology, and that relevant laws will need to be passed. Clearview AI was part of that pressure. "It was somewhere between striking and astonishing to see both how far they had advanced the technology and the broad sweep of the use they were marketing," he said. "And it just, to me, raises so sharply the question of what limits need to be put in place."

ONE OF THE activists' most resonant arguments against facial recognition technology has been that it is racist and does not work as well on some groups as others. But the window of time for that criticism to be effective is closing as top developers have focused on addressing the problem of biased algorithms, at least in test scenarios. One activist said privately that the advocacy community had "led with its chin" by focusing so much attention on a fixable problem with the technology, as opposed to the broader threats it would pose when perfected. That had provided facial recognition purveyors the opportunity to use greater accuracy across diverse groups as a justification to deploy the technology more widely.

Clearview did just that in February 2022 after members of Congress sent letters to five federal agencies asking them to stop using the company's product. The lawmakers, who included Washington representative Pramila Jayapal, who was born in India, and Massachusetts representative Ayanna Pressley, who is Black, expressed concern in their letters that "facial recognition technology like Clearview's poses unique threats to marginalized communities in ways that extend beyond the tools' inaccuracy issues. Communities of color are systematically subjected to over-policing, and the proliferation of biometric surveillance tools is, therefore, likely to disproportionately infringe upon the privacy of individuals in Black, Brown, and immigrant communities."

Ton-That issued a statement in response. Sidestepping the problem of overpolicing, he focused on the accusation of inaccuracy and pointed to the company's recent results from NIST, which was now running tests every few months on hundreds of facial recognition algorithms. At the end of 2021, after three years of its app's use by police in criminal investigations, Clearview had finally submitted its algorithm for third-party testing. In preparation for the evaluation, Clearview had brought back Terence Liu, the physicist from the University of Toledo who had helped Ton-That perfect the company's facial recognition algorithm years earlier. Ton-That was so nervous that he built a spider to crawl NIST's page every few seconds so that he would be alerted the instant the results were live.

To the surprise of some critics, Clearview ranked among the world's most accurate facial recognition companies, correctly matching two different mug shots of a person 99.76 percent of the time and someone's visa photo to a photo of them taken by a kiosk cam 99.7 percent of the time. Clearview debuted with the best algorithm among American companies, putting it on par with the heavyweights of the world, including Japan's NEC and NtechLab in Russia. The world's leader in facial recognition was SenseTime, the controversial vendor based in China, whose technology had been blacklisted by the U.S. government.

"Recent NIST testing has shown that Clearview AI's facial recognition algorithm shows no detectable racial bias," Ton-That declared. That wasn't entirely accurate. A close reading of NIST's "report card" on the company's algorithm showed clear demographic effects. While very accurate, Clearview's algorithm was more likely to confuse the faces of people born in Nigeria and Kenya than of those born in Poland. And Clearview was better at identifying men than women across the board.

Clearview had other critics in D.C. beyond activists and elected officials. By 2022, Alvaro Bedoya, Al Franken's former staffer and longtime critic of facial recognition, had been appointed a commissioner of the Federal Trade Commission, America's de facto privacy regulator. Bedoya said he planned to look at whether nonconsensual collection of people's biometric information was an unfair and deceptive business practice that potentially violated consumer protection laws.

"It's companies you've never heard of and don't have a relationship with that are scraping your face, your voice, the way you walk off of the internet," Bedoya said. He worried that companies stripping us of our anonymity as we move through the world will fundamentally change what it means to go "outside." "Do I want to live in a society where people can be identified secretly and at a distance by the government? I do not, and I think I'm not alone in that."

ABSENT ANY CONCRETE steps from lawmakers, privacy activists of a technical bent have come up with outlandish inventions to stop facial recognition technology. An eyewear maker in Michigan named Scott Urban has developed a line of reflective frames that bounce back visible and infrared light. When a surveillance camera films a person wearing the $168 Reflectacles, the reflected light blurs out the face. He also makes lenses that absorb infrared light to deter iris scanning. His customers include privacy enthusiasts, political activists, and card counters whose faces have been placed on casinos' watch lists. "People into their privacy are no longer shunned as loonies," Urban said.

Back in 2010, when the artist and technologist Adam Harvey was an NYU graduate student, he invented CV Dazzle—camouflage against facial recognition. It was a futuristic makeover involving funky checkerboard makeup and asymmetric hairstyles that covered key parts of the face. The dissonant, geometric designs worked against a face-detecting algorithm called Viola-Jones, which was widely used by facial recognition programs at the time.

Academics have called creations like these "privacy armor." But their protections are temporary, because surveillance merchants inevitably find a way to overcome them. As soon as those systems adopted a better algorithm, which they soon did, CV Dazzle no longer worked. To this day, journalists recommend Harvey's makeover as a way to subvert face surveillance, despite his tweets of protest.

Harvey, now based in Germany, conceived a different approach after hearing an intelligence officer remark that "an algorithm without data is useless." He reviewed hundreds of academic papers about facial recognition technology to see which data sets were being used to train and test their algorithms. He realized that the large community of developers was dependent on a handful of obscure public databases. They ranged from Duke MTMC, 2 million images of 2,700 students filmed walking around campus at Duke University, to MegaFace, more than 600,000 faces collected from photos posted publicly to the image-sharing site Flickr.

The MegaFace database had been created by University of Washington professors to run public performance tests of facial recognition algorithms, much like NIST's. Its collection of faces mined from wedding photos and kids' birthday parties had been downloaded more than six thousand times, by entities including Megvii and SenseTime, the two Chinese companies blacklisted by the U.S. government for human rights abuses; the military contractor Northrop Grumman; the Turkish police; a startup called ShopperTrak; and, on March 18, 2018, Hoan Ton-That of Smartcheckr.

After Harvey documented the existence of the databases, virtually unknown until then outside of a small computer vision research community, on a website called Exposing.ai, many were taken down, including MegaFace. That had been his goal. "I was interrupting the supply chain," he said. "Facial recognition doesn't work without people. If they collectively can refuse to have their data be part of the system, then it breaks."

Or they could do the opposite: They could break the system *with* their data. That was the tack that University of Chicago computer science professor Ben Zhao pursued. "Our goal is to make Clearview go away," Zhao said. Zhao and his students came up with a subversive software that they called Fawkes, in honor of Guy Fawkes, the English revolutionary whose arched brows, upturned mustache, and goatee adorn masks favored by protesters worldwide. Zhao's Fawkes software made pixel-level changes to a digital photograph that would confuse a facial recognition system later trying to match a person's face to that online photo. To the AI, it wouldn't look like the same person. It would work best if applied to at least half of all digital images of you, a challenging feat. But if Facebook, for example, deployed Fawkes on all of its users' photos, images scraped from the site without permission would no longer be a reliable way to identify someone.

UNTIL CLEARVIEW CAME along, Facebook had been the main target for activists and lawmakers concerned about facial recognition technology. Now that the landscape had changed, Facebook had a decision to make: publicly release the most recent version of the name-calling hat or not. The company was working on augmented reality (AR) glasses, which were essentially a wearable computer for the eyes, and some employees were wondering if these glasses would have the ability to identify people around you.

Andrew "Boz" Bosworth, a beefy, bald-headed Harvard classmate of Mark Zuckerberg who was soon to be named Facebook's chief technology officer, was in charge of the development of the glasses. During an all-hands meeting in early 2021, a concerned employee asked Boz, "When AR glasses are common, will people be able to mark their face as unsearchable?"

"Congratulations," Boz responded jauntily. "You are right on the money. This is the topic. I literally have a big privacy review today on this topic. I've got a conversation with Mark."

Meaning Zuckerberg.

"We've been working on this actually for years," he continued. He said he thought it would be possible to opt out of being searched, but that the bigger question was about whether Facebook would be able to release a face search at all, because facial recognition like that was illegal in some places, such as Illinois.

Boz described leaving facial recognition out of augmented reality glasses as a lost opportunity to augment human memory. He talked about the universal experience of going to a dinner party and seeing someone you know but failing to recall their name.

"We could put a little name tag on them," he said, with a short chuckle. "We could. We have that ability."

Years earlier, Boz himself had famously written an internal memo, titled "The Ugly," that said it was Facebook's job to connect people, no matter the outcome. It could help someone find love, which he described as "positive." Or it could cause loss of life because of a terrorist attack coordinated on Facebook's platform, which he deemed "negative." "The ugly truth is that we believe in connecting people so deeply that anything that allows us to connect more people more often is *de facto* good," he'd concluded.

Facial recognition technology would certainly connect people. While Facebook has since released smartglasses ($299 Ray-Bans with cameras, microphones, and a computer processor), true augmented reality glasses, with a digital overlay to surface information about the real world, are still in the pipeline, and it is still to be seen what they will be capable of.

Boz's remarks about the legality of putting virtual name tags on people may have sounded off-the-cuff, but in fact, they were the ripple on deep water. Employees from Facebook's artificial intelligence, policy, product, engineering, and legal teams had been asked by leadership, according to the company, to assess what Facebook should do about facial recognition technology. Facebook, at that time, had the world's largest global face biometric database, a creation so contentious that it had cost the company at least a billion dollars in legal costs and settlements alone.

The internal deliberation eventually took the form of a fifty-page memo, analyzing the pros and cons of maintaining this technology in the company's main product—with an emphasis on the cons. At the end of 2021, Facebook announced that it was pulling the plug on its facial recognition system; users would no longer be automatically recognized in photos and videos, and the company would delete the algorithmic maps it had made of more than a billion people's faces. Privacy advocates and activists applauded the decision. Rejecting a technology it had forced on its users a decade before was a weighty and symbolic move for Facebook, and it demonstrated just how toxic the idea of facial recognition had become. But it was also just that: symbolic.

Facebook wasn't getting rid of the algorithm that powered the system, just turning it off. "It's like putting the weapons in the armory," a former Facebook product manager said. "You can still go open the door anytime you want."

When I asked a Facebook spokesman whether the company might still put facial recognition into its augmented reality glasses one day, he refused to rule out the possibility.

CHAPTER 25

TECH ISSUES

On a Friday in October 2021, I experienced a future that may await us all. I was meeting with Hoan Ton-That to try out a product he'd once told me that Clearview had no plans to release.

We met at Lisa Linden's apartment on the Upper West Side, a shrine to antiquity decorated with early-American folk art, vintage radios and typewriters, a turn-of-the-century Graham telephone, and a candle snuffer from the 1800s. She prided herself on not being a tech person, despite representing a company on its cutting edge.

When I arrived, Ton-That was at peak tech geek, sitting at the dining room table, hunched over his laptop, looking very stressed out. He was wearing a sage green suit with a white belt and cream-colored shoes, and he had two sets of glasses on, a black chunky pair over more fashionable transparent prescription ones. He was alternating between squinting into the doubled-up glasses, typing on his computer, and looking at two different iPhones.

"There's something wrong with the internet," he complained to Linden. It was as if the historical relics in the room were exuding a mystical force against the future Ton-That was trying to usher in.

"Life stopped at about 1880 in this apartment," quipped Linden, who was dressed in a black suit with dangly diamond earrings. She told Ton-That that we should move down the hallway to a sitting room closer to the router in her bedroom. He said that wasn't the issue and spent more than an hour troubleshooting: plugging and unplugging things, checking connections, and reentering passwords.

While we waited, Ton-That told me that Clearview's database was growing at a rate of 75 million images per day. Previously, Clearview had, on average, found thirteen photos when a face was searched; now it was double that. Many of the new photos had come from Flickr.

"I started finding more photos of myself that I didn't know existed," Ton-That said and turned his phone toward me to show the surprising photos that had started appearing in his Clearview results: a photo from a decade earlier, of him in a dimly lit San Francisco bar, holding a drink; a teenage Ton-That, his hair short and shaggy, seated at a banquet table at a computer programming conference in Australia. His name was not on the photos. They were impossible to find in a Google word search. (I tried to find them later and couldn't.) Only a facial search like Clearview's would unearth them.

"It's a time machine. We invented it," Ton-That said. I saw no photos of him at the Republican National Convention or wearing a MAGA hat. Maybe they'd been crowded out—or maybe they'd been blocked from coming up.

During the course of my reporting this book, Ton-That snapped me with the Clearview app several times. Each time, the results were slightly different but always astounding: It would turn up my face in the background of strangers' photos, at a club, or walking down the street. Sometimes the photos would be from years back, from afar, and I could only be sure it was really me when I recognized the clothes I was wearing. Even when Ton-That took a photo of me wearing an N95 mask, covering a great deal of my face, the Clearview app still turned up photos of me.

During the search he did in Linden's apartment, the photo that captured my attention was from New Year's Eve 2006, spent with my sister and her college friends in New York City. I remembered being mildly depressed then, having just had a major breakup, and my weight was about twenty pounds heavier than it is now. I have few photos of myself online from that time, in part because I don't like how I looked then.

I didn't recall having seen the photo before, taken with a digital camera in the presmartphone era. The URL revealed that it had come from Flickr, posted, I later found out, by a friend my younger sister had fallen out of touch with. There were other photos of me there and many more of my sister. I was surprised to see that, because she is an extremely private person and has posted few photos of herself online. After I told her about this, she reached out to her friend and asked her to take the photos off the public web—but they would remain in Clearview's database.

Ton-That told me that Clearview was experimenting with new features made possible by artificial intelligence, such as making a blurry face clear and removing a person's Covid mask. He was also working on a new kind of search, not related to people's faces at all, of the *background* of a photo. He wanted investigators to be able to determine where a photo had been taken, to know that bricks in a wall, for example, were from a particular neighborhood in London. Human investigators already did that kind of analysis; he was sure he could train a computer to do it, too.

We talked about a recent report from the U.S. Government Accountability Office, a federal watchdog that had surveyed major agencies, including the FBI, the Department of Justice, the Department of Energy, the Secret Service, and the Department of Health and Human Services, about their use of facial recognition. Almost all of them had reported using it or planning to use it, with many reporting that they had tried Clearview. "You're becoming the Kleenex of facial recognition," I told Ton-That, the brand name that stands in for the generic.

That excited him, and he showed me a clip from a show that had aired the night before, a thriller based on a Dan Brown book. A character in the show calls a police friend for help identifying the face of someone who just shot at him. "Run it through Clearview. Get me a name," the guy says. "I owe you, brother."

Ton-That was giddy about it. That was his dream, to create a famous app, the kind that is casually name-dropped in a hot new TV show.*

Questions about Ton-That's time with the MAGA crowd and his and the company's association with Charles Johnson were off limits, as usual. Ton-That refused to talk about it, but he and Schwartz had finally found a way to dislodge Johnson from their company, and they had used me to do it. In early 2021, Schwartz filed an amendment to

*It was called *The Lost Symbol,* and it was canceled after one season.

Clearview's incorporation documents in Delaware. Any shareholder who "breaches any confidentiality obligations" could have their shares bought back at 20 percent of market value. The company knew that I had an article coming out in *The New York Times Magazine* that would reveal Johnson's role in the company and that it would include quotes from him. I found the new filing before the article published and asked Johnson if he was aware of it. He sounded shocked. "That's probably not good for me," he said.

After my story came out, the company's board of directors signed a legal document making the buyback official: It said that Johnson's "on the record" comments to *The New York Times Magazine* about his "indirect ownership stake" in Clearview had violated his duty of confidentiality. Under Clearview's new corporate policy, they were seizing Johnson's shares, worth nearly $1.5 million at the company's then valuation, and would pay him $286,579.69 for them. According to Johnson, Ton-That also called his ex-wife, asking to buy the shares she and their daughter held, but she refused.

After this news sunk in, Johnson acted blasé about it. He had moved from Texas to Virginia, had a new girlfriend, and had his hands full, he said, with a new startup called Traitwell, which would invite people to upload their DNA results from companies such as 23andMe and Ancestry, in order to "learn new and dangerous things about yourself." One of its offerings was a "Covid forecaster" to predict the severity of one's symptoms if infected. Johnson's true goal, though, was to build a genetic database that police could use to help solve crimes; this would ideally include everyone's DNA or at least that of someone related to them, which would be enough to make them findable in a genetic search. He said he planned to pay funeral homes to extract genetic samples from dead bodies. Johnson had a knack for being in the vanguard of dystopia. He swore he wouldn't cash a check from Clearview, planning instead to sue for what he felt he was truly owed.*

But this topic was verboten for Ton-That. Clearview's lawyer, Sam

*In March 2023, Johnson sued the company in the Southern District of New York, alleging breach of contract. A lawyer for the company said the lawsuit was without merit and that Clearview would defend itself "against these baseless claims."

Waxman, said they'd signed an NDA with Johnson when they had wound down Smartcheckr LLC years earlier, and no matter how much Johnson talked, they would not—a convenient way of avoiding an uncomfortable topic.

FINALLY, AFTER AN hour and a half, we moved into the sitting room, as Linden had recommended, and there the Wi-Fi did, magically, work. Ton-That was able to get the chunky black augmented reality glasses, made by a New York–based company called Vuzix, connected to the internet. As my colleagues and I had discovered when we first started looking into the company, Ton-That had designed Clearview's app so that it could run on the $999 glasses, allowing the wearer to look at a stranger from as far as ten feet away and find out who they were. It had been on the road map for years.

Ton-That turned toward me and tapped the glasses. "Ooooh. A hundred and seventy-six photos," he said. He was looking in my direction, but his eyes were focused on a small square of light on the right lens of the glasses.

"Aspen Ideas Festival. Kashmir Hill," he read. In the glasses, a photo had appeared from a conference where I had spoken in 2018, along with a link to the website where it had appeared and the page's label, which he had read verbatim.

Then he handed the glasses to me. I put them on. Though they looked ugly, they were lightweight and fit naturally. Ton-That said he had tried out other augmented reality glasses, but these had performed best. "They've got a new version coming," he said. "And they'll look cooler, more hipster."

Clearview would soon sign a contract with the U.S. Air Force to develop a prototype of the glasses that could be used on bases. Ton-That eventually wanted to develop a pair that could identify people at a distance, so that a soldier could make a decision about whether someone was dangerous or not when they were still fifty feet away.

When I looked at Ton-That through the glasses, a green circle appeared around his face. I tapped the touch pad at my right temple. A message came up on the square display that only I could see: "Searching . . ."

And then the square filled with photos of Ton-That, a caption beneath each one. I scrolled through them using the touch pad. I tapped to select one that read "Clearview CEO, Hoan Ton-That," and it showed me that it had come from Clearview's website.

I looked at Lisa Linden and searched her face, and forty-nine photos came up, including one at the engagement party for her niece and another with a client that she asked me not to include in the book, casually revealing just how intrusive a search for a face can be, even for people who work for the company.

I wanted to take the glasses outside to see how they worked on people I didn't actually know, but Ton-That said we couldn't, both because we needed a Wi-Fi connection and because someone might recognize him and realize immediately what the glasses were and what they could do. He wasn't ready for that level of controversy yet.

It didn't frighten me, though I knew it should. It was clear to me that people who own the glasses will inevitably have power over those without them. But there was a certain thrill to seeing it work, like a magic trick successfully performed, that taste of technical sweetness.

It was astonishing that Ton-That had gone from building banal Facebook apps to creating a world-changing software like this. "I couldn't have done it if I had to build it from scratch," he said, namedropping some of the researchers who had advanced computer vision and artificial intelligence. "I was standing on the shoulders of giants."

Those giants, computer scientists who had toiled in academic labs and Silicon Valley offices, had paved the way not just for Clearview but for future data-mining companies that may come for our voices, our expressions, our DNA, and our thoughts. They yearned to make computers ever more powerful, without reckoning with the full scope of the consequences. Now we have to live with the results.

After the demo was over, Ton-That sat down and took out his phone to check his email, which had piled up over the two-hour interview. As he started to get lost in the screen, Linden reprimanded him. "Hoan, stay focused," she said. "What's the most important thing?"

"Media," he replied, relaxing into his chair and putting down his phone. "Media."

Ton-That had compared himself at one point to Ignaz Semmelweis, a Hungarian doctor from the early 1800s who had tried to convince other doctors to wash their hands before attending to women in childbirth. Semmelweis published studies showing that hand washing reduced mortality rates among the women, but he couldn't explain why, so doctors refused to do it. Semmelweis died in a mental asylum, his advice unheeded. Hand washing wasn't embraced until the discovery of the relationship between germs and disease years later. Once a person has discovered something extraordinary, said Ton-That, "the real job is to sell it to the public and get people comfortable with it."

Ton-That remained fiercely proud of building an app that could link a person's face to their identity and their vast digital footprint. He was ready to duke it out with critics, to fight lawsuits, to challenge governments that deemed it illegal, and to give as many interviews as needed to convince people that in the end their faces were not solely their own. He again attributed the global resistance to "future shock."

"It's time for the world to catch up," he said. "And they will."

ACKNOWLEDGMENTS

First off, thank you to the people named and unnamed who spoke to me for this book, indulged my curiosities, gave me ideas and tips, spent hours or days or years talking to me, answered my questions, and otherwise helped me find answers.

I'm grateful to my colleagues at *The New York Times,* particularly Jennifer Valentino-DeVries, Aaron Krolik, Gabriel Dance, Susan Beachy, and Kitty Bennett, the reporters and researchers with whom I worked on the story that started it all. David Enrich and Nick Summers both wisely suggested, after one too many Clearview AI pitches from me, that I should probably write a book. Bill Wasik, Willy Staley, and Vauhini Vara at the magazine helped imagine the longer-form version of the story. Thank you to Kevin Roose, Sheera Frenkel, Cecilia Kang, and Mike Isaac for your advice on the book writing process, and to Rachel Dry, Ellen Pollock, and Carolyn Ryan for giving me the time to write it. The *Times* is an incredible place to be a journalist, and all the more rewarding when kind notes arrive from the masthead and from the publisher, A.G. Sulzberger.

Two people who were especially helpful in kindling my interest in facial recognition technology are Adam Harvey and Freddy Martinez. My colleague turned competitor turned colleague again, Ryan Mac, was very generous in sharing knowledge and insights. Lisa Linden and Jason Grosse went above and beyond in connecting me with sources I needed to speak with to make this a more satisfying book.

My agent, Adam Eaglin, discouraged me from pitching a book until I had truly great material; I am grateful because I loved working on this and with him. He and Chris Wellbelove helped put the proposal into the immensely capable hands of Hilary Redmon and Assallah Tahir, whose critiques and suggestions made the manuscript shine.

I can't thank Robert Boynton enough for encouraging me to pursue "The Not-So Private Parts" as my portfolio project at NYU's journalism school more than a decade ago, and for more recently connecting me with the indispensable researcher and fact-checker Anna Venarchik. She found crucial material and worked tremendously hard to make sure the book is factual and accurate. Any errors that remain are my own.

The invaluable feedback from early readers made this a better book and helped keep me going. My eternal gratitude to Trevor Timm, Monica Oriti, Maureen Taravella, Noah McCormack, Kate Shem, Dan Auerbach, Jonathan Auerbach, Chris, Ben Wizner, Radiance Chapman, Elie Mystal, and, again, because he more than deserves it, David Enrich.

The Internet Archive provided access to reference texts that were otherwise seemingly unobtainable, and I relied on the Wayback Machine countless times. Thank you, Brewster Kahle, for that essential digital library. The public records center MuckRock was another godsend.

I had hoped to make my way to Russia and China for this book, but the pandemic and a war got in the way of that. I am thankful to my hosts on the trips that I did take—Daniel Rivero in Miami and Liz Sabri in London—and to the Kinds for the lovely dinner party.

My friends and family encouraged and inspired me, particularly my mother, who is my most devoted reader. Dad, Mom, Scheherazade, Radiance, Keith, Coby, and Kincaid: Thank you for always being there when I need you. While I locked myself away in my office to write, my daughters were loved and entertained by Sara, Tess, Mo, Jim, and Angie. And of course my husband, Trevor, who supports my work and enhances my existence in so many ways that it would be impossible to summarize them here, so I'll just say I love you and thanks for all the coffees.

Ellev and Kaia, you are at the center of everything and are the joys of my life. You asked me many times over the last year whether I was done with the book yet. Thanks to everyone acknowledged here, I can finally answer in the affirmative.

A NOTE ON SOURCES

Your Face Belongs to Us is based on three years of research, trips to Miami and London, and interviews with more than 150 people. The idea for the project came out of a January 19, 2020, article that I wrote for *The New York Times* headlined "The Secretive Company That Might End Privacy as We Know It." I also fleshed out some of the ideas for the book while reporting a *New York Times Magazine* story, "Your Face Is Not Your Own," published on March 18, 2021.

Clearview AI and Hoan Ton-That, specifically, largely cooperated with my reporting for this book, for which I am grateful. Ton-That sat down with me over a dozen times to talk about his past and the company's history.

However, there were certain parts of his personal history that he declined to revisit, and there was a period of time that he and the company largely refused to answer questions about—the time they came to describe as the "Smartcheckr one" period: "There was a previous entity set up as an LLC in New York with a similar name, but it was not intended for the purpose of developing facial recognition technology, and it conducted no business," said Ton-That in a statement. "Smartcheckr Corp was incorporated in August 2017 to develop facial recognition technology with Richard Schwartz and Hoan Ton-That as the only shareholders. This corporate entity was later renamed Clearview AI, Inc. All intellectual property involving facial recognition was developed under Clearview AI, Inc."

The company maintains that the first Smartcheckr, which Charles Johnson was a part of, was an empty shell that did nothing, amounted to nothing, and had nothing to do with facial recognition. The company largely declined to answer questions about activities associated with Smartcheckr one, including the DeploraBall and the pitch to Hungary (which explicitly mentioned facial recognition technology).

Charles Johnson filled in many holes in the company's history, providing emails about face analysis and matching that were exchanged between Schwartz, Ton-That, and Johnson during the Smartcheckr one period and beyond. Johnson was generous with his time and helped me to understand periods of the company's development that no one else involved wanted to acknowledge. He was not the only person the company refused to discuss. They also largely declined to discuss "Veritas Strategic Partners," the political consulting business that did not involve Johnson and was created after the formation of the second Smartcheckr. However, Johnson, Holly Lynch, and other sources, named and unnamed, provided information and documents that undermine Clearview's claims that the Smartcheckr first established in New York wasn't part of a single continuum that led to Clearview.

No narrator of an experience you have not lived through yourself can be considered fully reliable, so I have endeavored to corroborate the events described in this book with multiple witnesses, contemporary communications, photos or video recordings, financial documents, corporate presentations, legal memos, diary entries, case files, and other government records. The endnotes specify when dialogue is based on a person's recollection, as opposed to transcripts, recordings, emails, or text messages. Any description of inner thoughts is based on my conversations with that person during the time of the interview cited in the endnotes.

Many of the pioneers of facial recognition technology are still alive, and I interviewed them directly. The major exception is Woodrow Bledsoe, of Panoramic Research, and so I relied on computer science papers he published; work by previous researchers, including the journalist Shaun Raviv in his *Wired* article "The Secret History of Facial Recognition"; and materials from Bledsoe's archives, some of which were graciously provided by Raviv, who spent extensive time combing through the collection at the University of Texas at Austin. I attempted to interview Helen Chan Wolf, Bledsoe's collaborator on the 1965 paper "A Man-Machine Facial Recognition System," but she declined, saying it had all happened too long ago.

NOTES

Prologue: The Tip

vii **a legal memo:** Paul Clement, memorandum re: Legal Implications of Clearview Technology, August 19, 2019. Obtained by Freddie Martinez in a public records request to the Atlanta Police Department.

vii **movies such as *Minority Report*:** Technically, *Minority Report,* released in 2002 and based on the 1956 short story of the same name by Philip K. Dick, features iris scanning, not facial recognition, but they belong in the same camp of technologies that aim to link a person's body to his or her name.

vii **police began experimenting:** Andy Newman, "Those Dimples May Be Digits," *New York Times,* May 3, 2001.

vii **largely proved disappointing:** John D. Sutter, "Why Facial Recognition Isn't Scary—Yet," CNN, July 9, 2010.

viii **"98.6 percent accuracy rate":** "Stop Searching, Start Solving," promotional brochure from Clearview AI. Obtained by Freddie Martinez, 2019.

viii ***Harvard Law Review* article:** Samuel D. Warren and Louis D. Brandeis, "The Right to Privacy," *Harvard Law Review,* December 15, 1890.

ix **a federal workshop:** "Face Facts," held by the Federal Trade Commission on December 8, 2011, videos and transcripts available online at https://www.ftc.gov/news-events/events/2011/12/face-facts-forum-facial-recognition-technology.

ix **Even Steve Jobs:** Fred Vogelstein, "And Then Steve Said, 'Let There Be an iPhone,'" *New York Times Magazine,* October 4, 2013.

ix **aren't quite there yet:** The blood-testing company Theranos has become the canonical example of this. The startup claimed it could perform hundreds of tests on a single drop of blood, which was not remotely true. John Carreyrou, *Bad Blood: Secrets and Lies in a Silicon Valley Startup* (New York: Alfred A. Knopf, 2018).

x **The company's online presence:** Clearview.ai, December 5, 2019, via Wayback Machine.

xi **Business filings:** Certificate of Incorporation of Smartchecker Corp, Inc., August 3, 2017; Certificate of Amendment of Certificate of Incorporation of Smartchecker Corp, Inc., June 15, 2018; State of Delaware annual franchise tax report for Clearview AI, Inc., March 26, 2019.

xii **"I love it":** Author's interview with Nick Ferrara for *The New York Times,* 2019.

xiv **The only time it didn't:** Author's interview with Daniel Zientek for *The New York Times,* 2019.

xv **"Oh, Clearview's lawyers said":** Author's interview with David Scalzo and Terrance Berland for *The New York Times,* 2019.

xvi **"endless":** Author's interview with Eric Goldman for *The New York Times,* 2019.

xvii **Minutes later, his phone rang:** Author's interviews in 2019 for *The New York Times* and in 2021 with an investigator in Texas who participated on the condition that his name not be used.

PART I: THE FACE RACE

Chapter 1: A Strange Kind of Love

3 **Hoan Ton-That did not want:** Author's interview with Charles "Chuck" Johnson, 2021.

3 **Ton-That had always followed:** The description of Hoan Ton-That's early life is based on author's interviews with Quynh-Du Ton-That, his father, 2020; Sam Conway, a childhood friend, 2021; and Hoan Ton-That, 2021.

4 **For the first time:** Michael Arrington, "Facebook Launches Facebook Platform; They Are the Anti-MySpace," TechCrunch, May 24, 2007.

4 **The epitome of those:** Austin Carr, "'FarmVille' Maker Zynga Grows a $1B IPO in Facebook's Fertile Earth," Fast Company, July 1, 2011.

5 **"It's the craziest thing":** Ton-That's recollection of his conversation with Naval Ravikant in interview with the author, 2020.

5 **"You see a lot of this":** The description of Ton-That's time in San Francisco is based on author's interviews with him, 2020–2022.

5 **he got married:** Certificate of marriage, retrieved from San Francisco City Hall.

5 **More than 6 million Facebook users:** Owen Thomas, "Was an 'Anarcho-Transexual Afro-Chicano' Behind the IM Worm?," Gawker, February 25, 2009.

6 **Cambridge Analytica scandal:** Cambridge Analytica was a British political consulting company at the center of a huge privacy scandal in 2018. The company had gotten its hands on the "likes" and profile information of millions of Facebook users by hiring a university researcher who claimed to be collecting it for academic purposes. Because Cambridge Analytica had worked for presidential candidate Donald Trump in 2016 and for Leave.EU during the Brexit vote, people became convinced that its secret insight into the inner lives of Facebook users had helped both conservative campaigns win.

6 **Pavlovian game:** Erica Sadden, "Expando for iPhone: Filler with Obnoxious Upselling," Ars Technica, January 27, 2009.

7 **"Hey, check out this video":** Eric Eldon, "ViddyHo Gives GTalk Users a Case of the Worms," VentureBeat, February 25, 2009.

7 **"Fast-Spreading Phishing Scam":** Jenna Wortham, "Fast-Spreading Phishing Scam Hits Gmail Users," *New York Times,* February 24, 2009.

7 **sucking a lollipop:** Thomas, "Was an 'Anarcho-Transexual Afro-Chicano' Behind the IM Worm?"

8 **Valleywag follow-up post:** Owen Thomas, "'Anarcho-Transexual' Hacker Returns with New Scam Site," Gawker, March 10, 2009.

8 **An FAQ:** Fastforwarded.com, March 13, 2009, via Wayback Machine.

9 **a handful of forgettable apps:** Cam-Hoan Ton-That, publisher summary page, Sensor Tower, retrieved in December 2019. Hoan Ton-That went by Cam-Hoan professionally early on.

9 **a new iPhone developer tool from Apple:** Benjamin Mayo, "iOS 7 to Let Developers Detect Blinking and Smiling in Photos," 9to5Mac, June 25, 2013.

9 **The promo art:** "Everyone" app description page, AppAdvice, retrieved in December 2019.

9 **Ton-That befriended:** Author's interviews with acquaintances of Hoan Ton-That who spoke on the condition of anonymity, 2020–2021.

9 **"I'll stop using":** Hoan Ton-That Twitter page from January 2016, retrieved via Wayback Machine.

10 **an opportunity to launder money:** After Bill Clinton adviser John Podesta's inbox was hacked, WikiLeaks published the contents online, including an email from an aide to Clinton, who was seemingly disgruntled about public remarks by his daughter, Chelsea, and wrote that she should quiet down or face a possible "investigation into her getting paid for campaigning, using foundation resources for her wedding and life for a decade." Based on the email, some media outlets ran stories saying that the Clinton Foundation had paid for Chelsea Clinton's wedding, which Bill Clinton denied. Veronica Stracqualursi, "Bill Clinton: Accusation That Foundation Paid for Daughter's Wedding a 'Personal Insult,'" CNN, January 14, 2018.

10 **race and intelligence:** In 2005, some academics published a paper claiming that Ashkenazi Jews, descendants of Jewish settlers who lived along the Rhine River in the Middle Ages, were more intelligent than other ethnic groups. It cited the prevalence of Jewish people receiving the Nobel Prize and questionable old IQ test research (all while ignoring a huge body of work suggesting that IQ tests are flawed indicators of comparative intelligence). The most trenchant criticism of the paper came from the political journalist Matt Yglesias, who said that people who praise Jewish genes typically have "an anti-Black agenda" and "see it as a useful entry point into race science." Gregory Cochran, Jason Hardy, and Henry Harpending, "Natural History of Ashkenazi Intelligence," *Journal of Biosocial Science* 38, no. 5 (2006): 659–93; Matthew Yglesias, "The Controversy over Bret Stephens's Jewish Genius Column, Explained," Vox, December 30, 2019.

10 **"There are too many":** The number of immigrants in the United States has been on the rise for decades. In 2016, Pew Research estimated that there were nearly 44 million foreign-born people in the country, accounting for 13.5 percent of the population, a new record. Jynnah Radford and Abby Budiman, "2016, Foreign-Born Population in the United States Statistical Portrait," Pew Research Center, September 14, 2018.

10 **One of Ton-That's best friends:** Author's interview with Gustaf Alströmer, 2021.

11 **Johnson's name is rarely invoked:** Caitlin Dewey, "Charles Johnson, One of the Internet's Most Infamous Trolls, Has Finally Been Banned from Twitter," *Washington Post,* May 26, 2015.

11 **described himself:** Much of the information in this section is from author's interviews with Charles Johnson, 2020–2021.

11 **He had written a book:** Charles C. Johnson, *Why Coolidge Matters: Leadership Lessons from America's Most Underrated President* (New York: Encounter Books, 2013).

11 **He had gained notoriety:** "Media Outlets Construct Michael Brown's (Thug) Life After Death," *Ebony,* September 4, 2014.

11 **GotNews got views:** "We had 3 or 4 million page views a month." Author's interview with Charles Johnson, 2021.

11 **derided by the mainstream media:** David Carr, "Sowing Mayhem, One Click at a Time," *New York Times,* December 14, 2014.

11 **and by Gawker:** J. K. Trotter, "What Is Chuck Johnson, and Why? The Web's Worst Journalist, Explained," Gawker, December 9, 2014.

11 **a *Rolling Stone* article:** The article, published in November 2014, described a gang rape at a fraternity on the UVA campus that had led to no consequences for the alleged rapists. *Rolling Stone* later retracted the story and paid the fraternity $1.65 million to settle a defamation lawsuit. Sabrina Rubin Erdely, "A Rape on Campus," *Rolling Stone,* November 2014.

11 **posted a photo:** Catherine Thompson, "How One Conservative Writer Mistook a Viral Photo for Rolling Stone's 'Jackie,'" Talking Points Memo, December 19, 2014.

12 **asked to be added:** This message, "add to Slack group," sent by Hoan Ton-That on May 16, 2016, and other messages described in this section were provided to author by Charles Johnson.

12 **"I loved Hoan":** Author's interview with Charles Johnson, 2021.

13 **Though it was generally regarded:** A nonexhaustive list: Kate Crawford, *Atlas of AI: Power, Politics, and the Planetary Costs of Artificial Intelligence* (New Haven: Yale University Press, 2021); Sahil Chinoy, "The Racist History Behind Facial Recognition," *New York Times,* July 10, 2019; "Eugenics and Scientific Racism," National Human Genome Research Institute.

13 **"I'll bring my guitar":** Email thread between Hoan Ton-That and Charles Johnson, July 13, 2016. On file with author.

13 **the philosopher-lawyer:** Thiel's biography comes primarily from Max Chafkin, *The Contrarian: Peter Thiel and Silicon Valley's Pursuit of Power* (New York: Penguin Press, 2021).

13 **who believed that:** Peter Thiel's views as described here are based on an essay he wrote: Peter Thiel, "The Education of a Libertarian," *Cato Unbound,* April 13, 2009.

13 **A gender studies professor:** Author's interview with Airbnb host for *The New York Times Magazine,* 2021. She agreed to talk on the condition that she not be named.

14 **frequently serenaded his friends:** The events at the Republican National

Convention were recounted by Charles Johnson in an interview with the author, 2021, and confirmed with photos and videos he provided, as well as a video posted to YouTube that shows Ton-That laughing after Johnson has a confrontation with a journalist on the street. GotNews, "Jamie Weinstein Assaults Charles C. Johnson," YouTube, July 21, 2016.

14 **Huge photos of NBA superstar LeBron James:** Don Muret, "Quicken Loans Arena Transforms for Republican National Convention," *Sporting News,* July 18, 2016.

14 **The main topic:** This meeting and much of description of Ton-That and Johnson's activities at the Republican National Convention are based on author's interviews with Charles Johnson, 2021–2022. Ton-That did not want to discuss the event, nor his association with Johnson. His spokeswoman claimed that doing so would violate an NDA. Peter Thiel was contacted through his spokesperson to confirm details about the poolside meeting, but he did not respond.

14 **closely covered the travails:** Author's interview with Owen Thomas, 2021.

14 **lost billions of dollars:** Ryan Tate, "A Facebook Billionaire's Big Dumb Failure." Gawker, August 16, 2011.

14 **"I think they should":** Connie Loizos, "Peter Thiel on Valleywag; It's the 'Silicon Valley Equivalent of Al Qaeda,'" *Venture Capital Journal,* May 18, 2009.

14 **The billionaire had been secretly:** Andrew Ross Sorkin, "Peter Thiel, Tech Billionaire, Reveals Secret War with Gawker," *New York Times,* May 25, 2016.

15 **it would settle:** Jonathan Randles, "Right-Wing Blogger Charles Johnson Settles with Gawker Media," *Wall Street Journal,* August 16, 2018.

15 **seemed to lend credence:** In his 2009 *Cato Unbound* essay, Thiel wrote that he stood against the "inevitability of death of every individual." In 2014, Thiel told Bloomberg TV that he was taking human-growth hormones in order to live to 120 years old. In 2015, he said he was interested in the life extension possibilities of blood transfusions from the young to the old, a process called parabiosis. Jeff Bercovici, "Peter Thiel Is Very, Very Interested in Young People's Blood," *Inc.,* August 1, 2016.

16 **"If you're a weird person":** Author's interview with Charles Johnson, 2021.

16 **"Do you want":** Conversation as recalled by Johnson in interviews with the author, 2020–2021.

Chapter 2: The Roots (350 B.C.–1880s)

17 **Aristotle declared:** Aristotle, *The History of Animals,* 350 BCE, book 1, parts 8–15.

17 **Galton, born in 1822:** Details of Galton's life come from an assortment of texts, including Karl Pearson, *The Life, Letters and Labours of Francis Galton* (Cambridge, UK: Cambridge University Press, 2011); Martin Brookes, *Extreme Measures: The Dark Visions and Bright Ideas of Francis Galton* (New

York: Bloomsbury, 2004); and Galton's writings, including *Narrative of an Explorer in Tropical South Africa* (London: John Murray, 1853) and *Memories of My Life* (London: Methuen, 1908).

18 **His older cousin was:** Much of the description of the relationship between Galton and Darwin comes from these articles: Raymond E. Fancher, "Scientific Cousins: The Relationship Between Charles Darwin and Francis Galton," *American Psychologist* 64, no. 2 (2009): 84–92; Nicholas W. Gillham, "Cousins: Charles Darwin, Sir Francis Galton and the Birth of Eugenics," *Significance* 6, no. 3 (August 24, 2009): 132–35.

18 **nearly rejected:** Charles Darwin, *The Life and Letters of Charles Darwin,* edited by Francis Darwin (London: John Murray, 1887), 59.

18 **a seven-page report:** Cornelius Donovan, "Phrenological Report on Francis Galton's Character," April 1849, Galton Papers, held at UCL Special Collections and Archives and digitized by the Wellcome Collection, https://wellcomecollection.org/works/mu8bwnff.

19 **"What labours and dangers":** Charles Darwin to Francis Galton, July 24, 1853, "Correspondence between Charles Darwin and Francis Galton," Sir Francis Galton FRS, https://galton.org/letters/darwin/correspondence.htm.

19 **he looked at the pedigrees:** This section is a summation of Galton's writings, including a two-part article—"Hereditary Talent and Character," *Macmillan's Magazine,* June 1865, 157–66, and August 1865, 318–27—and two books: *Hereditary Genius* (London: Macmillan, 1869); and *English Men of Science* (London: Macmillan, 1874).

19 **superimposed photographs of criminals:** Francis Galton, "Composite Portraits, Made by Combining Those of Many Different Persons into a Single Resultant Figure," *Journal of the Anthropological Institute of Great Britain and Ireland* 8 (1879): 132–44; Francis Galton, "Composite Portraits of Criminal Types," 1877, Sir Francis Galton FRS, https://galton.org/essays/1870-1879/galton-1879-jaigi-composite-portraits.pdf.

20 **He coined the term:** "We greatly want a brief word to express the science of improving stock, which is by no means confined to questions of judicious mating, but which, especially in the case of man, takes cognisance of all influences that tend in however remote a degree to give to the more suitable races or strains of blood a better chance of prevailing speedily over the less suitable than they otherwise would have had. The word *eugenics* would sufficiently express the idea," Galton wrote in *Inquiries into Human Faculty and Its Development* (London: Macmillan, 1883), 25.

20 **Charles Darwin wrote:** Charles Darwin to Francis Galton, December 23, 1869, Darwin Correspondence Project, University of Cambridge, https://www.darwinproject.ac.uk/letter/DCP-LETT-7032.xml.

20 **Reviews of the book were mixed:** Alfred Russel Wallace, review of *Hereditary Genius, Nature,* March 17, 1870. See also Emel Aileen Gökyiğit, "The Reception of Francis Galton's 'Hereditary Genius' in the Victorian Periodical Press," *Journal of the History of Biology* 27, no. 2 (1994): 215–40.

21 **Bertillon was tall and haughty:** The descriptions of Bertillon in this section

come from several sources: Richard Farebrother and Julian Champkin, "Alphonse Bertillon and the Measure of Man: More Expert than Sherlock Holmes," *Significance* 11, no. 2 (2014); Ida M. Tarbell, "Identification of Criminals: The Scientific Method in Use in France," *McClure's Magazine* 2, no. 4 (1894): 355–69; and a biography of Bertillon from "Visible Proofs: Forensic Views of the Body," exhibition at the National Library of Medicine, closed on February 25, 2008, but archived online.

21 **France had stopped branding:** It was outlawed in 1832, according to Dorothy and Thomas Hoobler, *The Crimes of Paris: A True Story of Murder, Theft, and Detection* (New York: Little, Brown, 2009).

22 **he was sorely disappointed:** Francis Galton, *Finger Prints* (London: Macmillan, 1892).

22 **distinct physical and mental characteristics:** Gina Lombroso Ferrero, *Criminal Man: According to the Classification of Cesare Lombroso* (New York: G. P. Putnam's Sons, 1911).

23 **police asked Lombroso:** Cesare Lombroso, *Crime: Its Causes and Remedies* (London: William Heinemann, 1911).

23 **"Lombroso has demonstrated":** Charles A. Ellwood, "Lombroso's Theory of Crime," *Journal of the American Institute of Criminal Law and Criminology* 2, no. 5 (1912): 716–23.

23 **an embrace of statistics:** Daniel J. Boorstin, *The Americans: The Democratic Experience* (New York: Vintage, 1974).

23 **Railways employed a rudimentary:** Information about the railways comes from two sources: Jim Strickland, "Hollerith and the 'Punched Photograph,'" Computer History Museum Volunteer Information Exchange, http://ed-thelen.org/comp-hist/images/VIE_04_003.pdf; and "The Punch Photograph," *The Railway News* 48, no. 1234 (1887): 360.

24 **Much complaint has been made:** "The Punch Photograph."

24 **inspired one important traveler:** In an August 1919 letter, Hollerith wrote of his inspiration, "I was traveling in the West and I had a ticket with what I think was called a punch photograph . . . [T]he conductor . . . punched out a description of the individual, as light hair, dark eyes, large nose, etc. So you see, I only made a punch photograph of each person." Quoted in Geoffrey D. Austrian, *Herman Hollerith: Forgotten Giant of Information Processing* (New York: Columbia University Press, 1982), 15.

24 **thought the "punch" method:** "The Hollerith Machine," United States Census Bureau, https://www.census.gov/history/www/innovations/technology/the_hollerith_tabulator.html; Leon Edgar Truesdell, *The Development of Punch Card Tabulation in the Bureau of the Census, 1890–1940* (Washington, D.C.: U.S. Government Printing Office, 1965).

25 **Hooton took the measurements:** Earnest Albert Hooton, *Crime and the Man* (Cambridge, Mass.: Harvard University Press, 1939).

25 **gets less attention:** Alexandra Minna Stern, "Forced Sterilization Policies in the US Targeted Minorities and Those with Disabilities—and Lasted into the 21st Century," The Conversation, August 26, 2020; Sanjana Manjeshwar,

"America's Forgotten History of Forced Sterilization," *Berkeley Political Review,* November 4, 2020.

26 **a chilling 1927 Supreme Court ruling:** *Buck v. Bell,* Supreme Court decision, May 2, 1927. See also Adam Cohen, *Imbeciles: The Supreme Court, American Eugenics, and the Sterilization of Carrie Buck* (New York: Penguin Books, 2017).

26 **"highly unorthodox":** Robert K. Merton and M. F. Ashley-Montagu, "Crime and the Anthropologist," *American Anthropologist* 42, no. 3 (1940): 384–408.

Chapter 3: "Fatface Is Real"

27 **Schwartz grew up:** Details of Schwartz's life are based on news coverage and author's interviews with Richard Schwartz, 2021. Any direct quotes are from author's interview.

27 **Robert Moses, a polarizing:** Ashish Valentine, "'The Wrong Complexion for Protection.' How Race Shaped America's Roadways and Cities," NPR, July 5, 2020.

28 **lucrative consulting business:** "On February 11, 1997, at age 38, Richard Schwartz announced he was leaving city government. The next day, he founded Opportunity America. . . . Business looks good so far: The tiny consulting firm managed to secure contracts worth about $5.5 million in a single month at the end of last year." Kathleen McGowan, "The Welfare Estate," City Limits, June 1, 1999.

28 **"system that once":** Ibid.

28 **New York City corruption scandal:** "Virginia-based Maximus, which has ties to a former mayoral adviser, had an unfair advantage over other groups vying for the lucrative welfare-to-work contracts, Manhattan Supreme Court Justice William Davis ruled." Kirsten Danis, "Judge Sides with Hevesi on Workfare Bids," *New York Post,* April 14, 2001; Bob Herbert, "Contracts for Cronies," *New York Times,* March 27, 2000.

28 **The controversy eventually blew over:** Eric Lipton, "Rejecting Favoritism Claim, Court Upholds a City Welfare Contract," *New York Times,* October 25, 2000; Michael Cooper, "Disputed Pacts for Welfare Will Just Die," *New York Times,* October 4, 2002.

28 **sold his firm:** An article about the purchase—Tom Robbins, "A Maximus Postscript," *Village Voice,* June 17, 2003—said that the price was $780,000, but in a Form 10-Q filed with the SEC for the quarter ending in June 2001, Maximus said that the purchase price was $825,000.

28 **media reporters who were perplexed:** "For city reporters accustomed to getting the big blow-off from the former Mayoral aide and workfare consultant, Mr. Schwartz's appointment was like hearing that the school bully got picked to be hall monitor." Gabriel Snyder, "How Welfare King Richard Schwartz Landed at the *Daily News,*" *Observer,* March 12, 2001. A former colleague was quoted putting Schwartz down: "In the world of ass kissers, Richard wins the Academy Award, the Tony, the Nobel Prize, the Pulitzer." Andrew Kirtzman, "Big Apple Polisher," *New York,* March 19, 2001.

28 **go into the behind-the-scenes world:** Author's interview with Ken Frydman, 2021.

29 **They met under a bridge:** Author's interview with Charles Johnson, 2021.

29 **"You should meet in NYC ASAP!":** Charles Johnson, email to Hoan Ton-That and Richard Schwartz, July 23, 2016. Provided by Johnson, on file with author.

29 **They soon did:** The description of Johnson's meeting with Schwartz and Johnson's introduction of Ton-That and Schwartz are based on author's interviews with Johnson, 2021, and emails provided by him.

29 **they hit it off:** The description of Ton-That and Schwartz's meeting is based on author's interviews with Ton-That, 2020, and Schwartz, 2021.

30 **people protested:** Manuel Perez-Rivas, "Koch Suspends Park Curfew Following Bloody Clash in Tompkins Square," *Newsday,* August 8, 1988; Robert D. McFadden, "Park Curfew Protest Erupts into a Battle and 38 Are Injured," *New York Times,* August 8, 1988.

30 **closed for a year:** John Kifner, "New York Closes Park to Homeless," *New York Times,* June 4, 1991.

31 **emailing one another frequently:** The descriptions of the group's emails and plans at the time are based on interviews with Charles Johnson, emails provided by him, and company presentations created around that time.

31 **guess people's sexuality:** Psychology researchers at Tufts University released two studies on this in 2008: Nicholas O. Rule and Nalini Ambady, "Brief Exposures: Male Sexual Orientation Is Accurately Perceived at 50 Ms," *Journal of Experimental Social Psychology* 44, no. 4 (2008): 1100–05; and Nicholas O. Rule et al., "Accuracy and Awareness in the Perception and Categorization of Male Sexual Orientation," *Journal of Personality and Social Psychology* 95, no. 5 (2008): 1019–28. In 2015, researchers at the University of Wisconsin–Madison released a responsive study challenging the "gaydar myth."

31 **same experiment with AI:** Yilun Wang and Michal Kosinski, "Deep Neural Networks Are More Accurate than Humans at Detecting Sexual Orientation from Facial Images," *Journal of Personality and Social Psychology* 114, no. 2 (2018): 246–57.

31 **effectively dismantled:** Mike Moffitt, "Gaydar Is Probably Not Real, After All," SFGATE, September 8, 2015; Alan Burdick, "The A.I. 'Gaydar' Study and the Real Dangers of Big Data," *The New Yorker,* September 15, 2017; Carl Bergstrom and Jevin West, "Machine Learning About Sexual Orientation?," Calling Bullshit, September 19, 2017.

31 **predict a "criminal face":** Xiaolin Wu and Xi Zhang, "Automated Inference on Criminality Using Face Images," arXiv, November 13, 2016.

31 **dueling American teams:** A press release from Harrisburg University in 2020 announced a paper titled "A Deep Neural Network Model to Predict Criminality Using Image Processing." The release stated that the software could predict criminality from a photo of a face "with 80 percent accuracy and with no racial bias." The announcement met with such a fierce backlash that it

was never released. See also Mahdi Hashemi and Margeret Hall, "RETRACTED ARTICLE: Criminal Tendency Detection from Facial Images and the Gender Bias Effect," *Journal of Big Data* 7 (2020): article 2, published January 7, 2020, retracted June 30, 2020. The researchers did not admit fault with their findings, saying instead that the reason they had withdrawn their paper was that they hadn't gotten approval from a university ethics board to use biometric data from the thousands of individuals whose photos they had used.

31 **thoroughly savaged:** Coalition for Critical Technology, "Abolish the #TechToPrisonPipeline," Medium, June 23, 2020.

31 **cropped from mug shots:** Kevin Bowyer, Michael C. King, and Walter Scheirer, "The 'Criminality from Face' Illusion," *IEEE Transactions on Technology and Society* 1, no. 4 (2020): 175–83. "Society at large should be very concerned if physiognomy makes a serious resurgence through computer vision," warned the researchers in their paper.

31 **"This is really dumb":** Hoan Ton-That, email to Richard Schwartz and Charles Johnson, October 17, 2016. Provided by Johnson, on file with author.

32 **"Fatface is real":** Hoan Ton-That, email to Richard Schwartz and Charles Johnson, September 12, 2016. Provided by Johnson, on file with author.

32 **predict a person's body mass index:** Lingyun Wen and Guodong Guo, "A Computational Approach to Body Mass Index Prediction from Face Images," *Image and Vision Computing* 31, no. 5 (2013): 392–400.

32 **dubious 2014 study:** Karel Kleisner, Veronika Chvátalová, and Jaroslav Flegr, "Perceived Intelligence Is Associated with Measured Intelligence in Men but Not Women," *PLoS* ONE, March 20, 2014.

32 **"Makes sense from evolution":** Ton-That, email to Richard Schwartz and Charles Johnson, September 12, 2016. Provided by Johnson, on file with author.

32 **"People putting crazy money":** Hoan Ton-That, email to Richard Schwartz and Charles Johnson, October 25, 2016. Provided by Johnson, on file with author.

33 **In a revealing 2015 hack:** Chris Isidore and David Goldman, "Ashley Madison Hackers Post Millions of Customer Names," CNN, August 19, 2015.

33 **"We could use email":** Hoan Ton-That, email to Richard Schwartz and Charles Johnson, November 22, 2016. Provided by Johnson, on file with author.

33 **When a couple of Danish researchers:** Brian Resnick, "Researchers Just Released Profile Data on 70,000 OkCupid Users Without Permission," Vox, May 12, 2016.

33 **"drug use and sexual preferences":** Hoan Ton-That, email to Richard Schwartz and Charles Johnson, November 23, 2016. Provided by Johnson, on file with author.

33 **believed in a link:** The Danish blogger, Emil Kirkegaard, writes on this topic often, as documented in a 2017 article about him: Judith Duportail, "Dans le laboratoire de la 'fake science'" ("In the Laboratory of 'Fake Science'"), *Le*

Temps, April 7, 2017. A selection of his work includes: Emil O. W. Kirkegaard, "Environmental Models of the Black-White IQ Gap," RPubs; and Emil O. W. Kirkegaard, "Expert Opinion on Race and Intelligence," Clear Language, Clear Mind, April 24, 2020.

33 **"He suggested scraping":** Hoan Ton-That, email to Richard Schwartz and Charles Johnson, November 28, 2016. Provided by Johnson, on file with author.

33 **Some of Ton-That's friends:** Author's interviews with multiple acquaintances who spoke on the condition that they not be named, 2020–2021.

33 **Johnson took him:** Video provided by Charles Johnson.

34 **"if our face matching":** Hoan Ton-That, email to Richard Schwartz and Charles Johnson, November 23, 2016. Provided by Johnson, on file with author.

34 **"Amazing! Can we beat":** Richard Schwartz, email to Hoan Ton-That and Charles Johnson, October 23, 2016. Provided by Johnson, on file with author.

34 **"This will be the future":** Hoan Ton-That, email to Richard Schwartz and Charles Johnson, November 23, 2016. Provided by Johnson, on file with author.

Chapter 4: If At First You Don't Succeed (1956–1991)

36 **Most Americans check:** "The New Normal: Phone Use Is up Nearly 4-Fold Since 2019," Asurion. This was based on two Asurion-sponsored market research surveys, the first in August 2019 of 1,998 U.S. smartphone users and the second in March 2022 of 1,965 U.S. adults.

36 **Periodic surveys:** Thomas Moller-Nielsen, "What Are Our Phones Doing to Us?," *Current Affairs,* August 14, 2022. See also Aaron Smith, "Chapter Two: Usage and Attitudes Toward Smartphones," Pew Research Center, April 1, 2015.

36 **create machines in our image:** John McCarthy et al., "A Proposal for the Dartmouth Summer Research Project on Artificial Intelligence," August 31, 1955, http://jmc.stanford.edu/articles/dartmouth/dartmouth.pdf.

36 **Many histories:** The development of AI as described in this section is based primarily on two books on the topic: Melanie Mitchell, *Artificial Intelligence: A Guide for Thinking Humans* (London: Picador, 2019) and Cade Metz, *Genius Makers: The Mavericks Who Brought AI to Google, Facebook, and the World* (New York: Dutton, 2021).

36 **the all-male roster:** The proposal called for a "2 month, 10 man study of artificial intelligence." A list of attendees kept by one participant, Trenchard More, listed thirty men who participated for varying amounts of time and was documented by Grace Solomonoff, "Ray Solomonoff and the Dartmouth Summer Research Project in Artificial Intelligence, 1956," http://raysolomonoff.com/dartmouth/dartray.pdf. The most famous photo of the male participants taken that summer lists Marvin Minksy's wife as the photographer, so there was at least one woman present, though she wasn't considered a participant.

36 **"An attempt will be made":** Ibid.

36 **half the money requested:** They wanted $13,000, but the Rockefeller Foundation gave them $7,500. Grant document RSM #51, Rockefeller Archive, December 28, 1955.

37 **two different ideas emerged:** "AI research split, perhaps even before 1956, into approaches based on imitating the nervous system and the engineering approach of looking at what problems the world presents to humans, animals, and machines attempting to achieve goals including survival." John McCarthy, "The Dartmouth Workshop—as Planned and as It Happened," October 30, 2006, Stanford University, http://www-formal.stanford.edu/jmc/slides/dartmouth/dartmouth/node1.html.

37 **almost single-handedly cutting off:** In 2006, the computer researcher Terry Sejnowski addressed Minsky during the Q and A at an AI conference. "There is a belief in the neural network community that you are the devil who was responsible for the neural network winter," Sejnowski said. He then asked him if he was the devil. "Yes, I am the devil," Minsky replied. Terrence J. Sejnowski, *The Deep Learning Revolution* (Cambridge, Mass.: MIT Press, 2018), 258.

37 **Woody Bledsoe:** The descriptions of Bledsoe's life and work are based primarily on two documents: Shaun Raviv, "The Secret History of Facial Recognition," *Wired,* January 21, 2020; and Michael Ballantyne, Robert S. Boyer, and Larry Hines, "Woody Bledsoe: His Life and Legacy," *AI Magazine* 17, no. 1 (1996): 7–20.

38 **a government patron:** A history of the CIA published in 1969 and obtained by Governmentattic.org discusses "two projects in pattern recognition" that the agency helped establish, including the "development of a man-machine system for facial recognition." "History of the Office of Research and Development, Volumes 1 through 6," declassified for release by the CIA on November 11, 2017.

38 **a line of research:** Shaun Raviv graciously shared digital versions of papers and materials from Woody Bledsoe's archive at the University of Texas, which I relied upon for this section.

38 **"a simplified face recognition machine":** W. W. Bledsoe, "Proposal for a Study to Determine the Feasibility of a Simplified Face Recognition Machine," January 30, 1963, Internet Archive, https://archive.org/details/firstfacialrecognitionresearch/FirstReport/page/n1/mode/2up.

38 **"millions or even billions":** W. W. Bledsoe, "The Model Method of Facial Recognition," August 12, 1964, Internet Archive, https://archive.org/details/firstfacialrecognitionresearch/FirstReport/page/n21/mode/2up.

38 **Bledsoe's proposal appears:** When asked for documents detailing the nature of the King-Hurley Research Group's relationship with the CIA, the agency told one public records requester that it could neither confirm nor deny the group's existence, a reply known as a Glomar response. Justine Lange, "Relationship of King-Hurley Research Group and CIA," MuckRock, April 24, 2014. When I sent a public records request to the CIA in 2021 asking for in-

formation or records related to Woody Bledsoe, I got a similar response. The agency does not officially acknowledge funding his work, but King-Hurley's role as a CIA money front from 1955 to 1969 has been documented in Bill Richards, "Firm's Suits Against CIA Shed Light on Clandestine Air Force," *Washington Post,* May 12, 1978; and Christopher Robbins, *The Invisible Air Force: The True Story of the Central Intelligence Agency's Secret Airlines* (London: Macmillan, 1981).

39 **a persistent identifier:** They called it a "set of 'standard' distances." W. W. Bledsoe and Helen Chan, "A Man-Machine Facial Recognition System: Some Preliminary Results," Panoramic Research, Inc., 1965. Available at the University of Texas Archive.

39 **"The important question":** Ibid.

39 **In Alabama:** "NAACP: A Century in the Fight for Freedom," Library of Congress, www.loc.gov/exhibits/naacp/the-civil-rights-era.html#obj13.

39 **FBI agents wiretapped King's home:** Tony Capaccio, "MLK's Speech Attracted FBI's Intense Attention," *Washington Post,* August 27, 2013; Beverly Gage, "What an Uncensored Letter to M.L.K. Reveals," *New York Times,* November 11, 2014; Sarah Pruitt, "Why the FBI Saw Martin Luther King Jr. as a Communist Threat," History, June 24, 2021.

39 **The most significant leap forward:** This is based on interviews with many technologists who specialize in computer vision. Kanade's 1977 paper was widely recognized as the most significant development in the field of facial recognition since Bledsoe's work. Takeo Kanade, *Computer Recognition of Human Faces* (Basel: Birkhäuser Verlag, 1977).

40 **The 1970 World Exposition:** Descriptions are based in part on a convention guidebook made available online by the U.S. Department of State and on "The 1970 Osaka World's Fair Was Something Else," Messynessy, February 12, 2020.

40 **Another popular exhibit:** "The History of NEC Technical Journal," NEC, https://www.nec.com/en/global/techrep/journal/history/index.html; author's interviews with Takeo Kanade, 2021.

41 **They were stored:** Much of this section is based on author's interviews with Takeo Kanade, 2021.

41 **"If you can process":** As recalled by Takeo Kanade in interview with the author, 2021.

42 **And then along came Matthew Turk:** Much of this section is based on the author's interviews with Matthew Turk, 2020–2021.

42 **paid at least $17 million:** "Engineers at Martin Marietta Put Their $17-Million Experiment in . . . ," UPI, July 12, 1986. David Ake is listed as the author of the story, but he is a photographer who would not have written it. He might have been assigned to take photos at the event, but he could not recall having done so.

42 **DARPA had tapped:** DARPA, *DARPA, 1958–2018,* September 5, 2018, https://www.darpa.mil/attachments/DARAPA60_publication-no-ads.pdf.

43 **According to one history:** Alex Roland and Philip Shiman, *Strategic Comput-*

ing: DARPA and the Quest for Machine Intelligence, 1983–1993 (Cambridge, Mass.: MIT Press, 2002), 246.

43 **remarkable, and ongoing:** A 1983 report from DARPA said that one of the purposes of the Strategic Computing Initiative was "to build up a base of engineers and systems builders familiar with computer science and machine intelligence technology." DARPA, *Strategic Computing: New-Generation Computing Technology: A Strategic Plan for Its Development and Application to Critical Problems in Defense,* October 28, 1983, 62. https://apps.dtic.mil/sti/pdfs/ADA141982.pdf.

43 **the MIT Media Lab:** In 2019, it was revealed that the Media Lab had accepted donations from the investor and convicted child sex offender Jeffrey Epstein. The deceased Minsky became part of the scandal after a victim of Epstein testified that she had been directed to have sex with the MIT professor when she was seventeen and Minsky was seventy-three. Joi Ito, "My Apology Regarding Jeffrey Epstein," MIT, August 15, 2019, https://www.media.mit.edu/posts/my-apology-regarding-jeffrey-epstein/; Russell Brandom, "AI Pioneer Accused of Having Sex with Trafficking Victim on Jeffrey Epstein's Island," The Verge, August 9, 2019.

43 **"invent the future":** Edward Dolnick, "Inventing the Future," *New York Times,* August 23, 1987.

43 **His main accomplishment:** "MIT Artificial Intelligence Laboratory Robotic Arm (Minsky Arm)," MIT Museum, https://webmuseum.mit.edu/detail.php?module=objects&type=popular&kv=67020.

44 **had just deemed practically impossible:** "No one has yet been able to build vision machines that approach our human ability to distinguish faces from other objects—or even to distinguish dogs from cats," Minsky had written in *The Society of the Mind,* his then-new book speculating on how the human brain worked, based in part on his experiences trying to get computers to simulate thought. Marvin Minsky, *The Society of Mind* (New York: Simon & Schuster, 1988), 312.

44 **"Face recognition was thought of":** Sandy Pentland, email exchange with the author, 2021.

44 **a new computer technique:** L. Sirovich and M. Kirby, "Low-Dimensional Procedure for the Characterization of Human Faces," *Journal of the Optical Society of America A* 4, no. 3 (1987): 519–24.

44 **Arbitron wanted:** Author's interview with Matthew Turk. See also James Roman, *Love, Light, and a Dream: Television's Past, Present, and Future* (Westport, Conn.: Praeger, 1998).

44 **who was in the room and when:** In a foreshadowing of the pervasive web-based tracking to come with the advent of computers and smartphones, Arbitron launched an initiative called "ScanAmerica" in 1987. It paid households up to $400 to wave a "scanner wand" over the product code of every item they bought. Arbitron wanted a 360-degree view of the ads a household saw and what they purchased—information it thought would be incredibly valuable to advertisers. Surprisingly, the journalists who covered this did not seem to

think it was completely insane; a news story at the time included video of a happy housewife in her kitchen dutifully scanning the UPC codes on everything she has just bought at the grocery store. Bill Carter, "And Arbitron Tries to Track Buying Habits," *New York Times,* November 4, 1991; See also "Can You Believe TV Ratings?," *The Cutting Edge,* 1992, YouTube, https://www.youtube.com/watch?v=N7LoUwOMZaA.

44 **"So that was really":** Author's interview with Matthew Turk, 2020.

44 **a baby, potentially within hours of birth:** One study—Elinor McKone, Kate Crookes, and Nancy Kanwisher, "The Cognitive and Neural Development of Face Recognition in Humans," in *The Cognitive Neurosciences,* 4th ed., edited by Michael S. Gazzaniga (Cambridge, Mass.: MIT Press, 2009), 467–82—says that it takes days for a baby to recognize its mother's face, while others—e.g., Francesca Simion and Elisa Di Giorgio, "Face Perception and Processing in Early Infancy: Inborn Predispositions and Developmental Changes," *Frontiers in Psychology* 6 (2015): 969—say that it takes hours.

45 **Neuroscientists thought:** This is based on author's interview with the cognitive neuroscientist Michael Tarr, 2022, and email exchange with MIT professor Nancy Kanwisher, 2022. Also Kanwisher's lecture "What You Can Learn from Studying Behavior" (video), nancysbraintalks, https://nancysbraintalks.mit.edu/video/what-you-can-learn-studying-behavior, September 8, 2014.

45 **one cognitive neuroscientist has said:** Author's interview with Michael Tarr, 2022. See also: Isabel Gauthier et al., "Expertise for Cars and Birds Recruits Brain Areas Involved in Face Recognition," *Nature Neuroscience* 3, no. 2 (February 2000).

45 **truly terrible at it:** Oliver Sacks, *The Man Who Mistook His Wife for a Hat* (New York: Summit Books, 1985).

45 **developed a method to:** Sirovich and Kirby, "Low-Dimensional Procedure for the Characterization of Human Faces."

46 **never wound up using the system:** Around the same time, Arbitron competitor Nielsen also announced plans to use facial recognition technology to keep track of TV viewers, working with researchers at Princeton University to develop it. These days, companies like this compete to see how much of your internet activity they can track. Nielsen is in that business in partnership with Arbitron, which it acquired in 2013. Bill Carter, "TV Viewers, Beware: Nielsen May Be Looking," *New York Times,* June 1, 1989.

46 **wrote a paper:** Matthew Turk and Alex Pentland, "Eigenfaces for Recognition," *Journal of Cognitive Neuroscience* 3, no. 1 (1991): 71–86.

47 **"This is old hat":** Turk's recollection in interview with the author, 2020. Author also corresponded by email with the computer vision pioneer Robert Haralick in 2021. He confirmed his skepticism about the novelty of the paper: "Approaches can be made to sound deep and sexy. Such is the case about *eigenfaces*. Principal Component Analysis has been around a long time." He did acknowledge that the *eigenfaces* technique helped computers with limited power process large images and "gave better answers than most of the competing techniques then."

47 **a life of its own:** Google Scholar says the paper has been cited 20,455 times. Retrieved February 2023.

47 **"technical sweetness":** Author's interview with Heather Douglas, 2022. See also Heather Douglas, *Science, Policy, and the Value-Free Ideal* (Pittsburgh: University of Pittsburgh Press, 2009).

48 **a documentary film crew came:** "Can You Believe TV Ratings?," *The Cutting Edge.*

48 **mistake them for animals:** In 2015, Google released an astounding tool for its Photos app that could automatically sort photos by what was in them. It could group all your photos containing bikes or all your pictures of food. But the new tool made an incredibly offensive mistake: A Black software engineer found that photos of him and a friend, who was a Black woman, had been labeled "gorillas." He tweeted a screenshot, writing "Google Photos, y'all fucked up." Google apologized, but rather than fixing the problem of Black people being identified as a different species, it blocked its image recognition algorithm from recognizing gorillas, along with monkeys, chimpanzees, and chimps. Tom Simonite, "When It Comes to Gorillas, Google Photos Remains Blind," *Wired,* January 11, 2018.

Chapter 5: A Disturbing Proposal

50 **On election day 2016:** Author's interview with Charles Johnson, 2021. Johnson also provided photos from the evening.

50 **"It was a sorry affair":** Unless otherwise noted, all quotes in this chapter from Charles Johnson are from author's interviews with him, 2021.

50 **despite investigative reporting to the contrary:** Suzanne Craig, "Trump's Empire: A Maze of Debts and Opaque Ties," *New York Times,* August 20, 2016; Tina Nguyen, "Is Donald Trump Not Really a Billionaire?," *Vanity Fair,* May 31, 2016; Russ Buettner and Charles V. Bagli, "Donald Trump's Business Decisions in '80s Nearly Led Him to Ruin," *New York Times,* October 3, 2016; Nicholas Confessore and Binyamin Applebaum, "How a Simple Tax Rule Let Donald Trump Turn a $916 Million Loss into a Plus," *New York Times,* October 3, 2016.

50 **later designated a hate group:** "Proud Boys," Southern Poverty Law Center, https://www.splcenter.org/fighting-hate/extremist-files/group/proud-boys.

50 **McInnes had formed:** Gavin McInnes, "Introducing: The Proud Boys," *Taki's Magazine,* September 15, 2016, https://archive.is/9Xs2K. Also Adam Leith Gollner, "The Secret History of Gavin McInnes," *Vanity Fair,* June 29, 2021.

50 **The Proud Boys' election night party:** Much of the description of the event is drawn from a promotional video created by the organizers. The credits on the nearly six-minute-long video identify the filmmaker Pawl Bazile as the director and Jack Buckby and Gavin McInnes as the producers. Charles Johnson and Hoan Ton-That are visible at the end of the video in the audience as it chants "USA! USA!" PawL BaZiLe, "Proud Boy Magazine 2016

Election Night Party Goes Full America!!!," YouTube, November 10, 2016, https://www.youtube.com/watch?v=hYYThd_J6U8.

50 **remark that half:** Katie Reilly, "Read Hillary Clinton's 'Basket of Deplorables' Remarks About Donald Trump Supporters," *Time,* September 10, 2016.

51 **"The Proud Boys will not be defied!":** Dialogue from BaZiLe, "Proud Boy Magazine 2016 Election Night Party Goes Full America!!!"

51 **"If Trump wins":** Ibid.

51 **"Light a cigarette!":** Ibid.

51 **hundreds of millions of dollars:** "Since 2001, when it made less than $1 million in federal contracts, Blackwater has received more than $1 billion in such contracts." "Blackwater's Rich Contracts," *New York Times,* October 3, 2007.

52 **After Blackwater mercenaries:** "Four Former Blackwater Employees Found Guilty of Charges in Fatal Nisur Square Shooting in Iraq," United States Department of Justice, October 22, 2014, https://www.justice.gov/opa/pr/four-former-blackwater-employees-found-guilty-charges-fatal-nisur-square-shooting-iraq.

52 **"Hoan and I":** Author's interview with Charles Johnson, 2021.

52 **went to a nearby diner:** Ibid. Erik Prince's trip to the diner with Johnson was also described by Prince himself in an article published a year later: "I ended up taking a bunch of people, Charles [Johnson] and a bunch of random people—one of the Baldwin brothers—out to a diner, and we had breakfast at like 5:30 in the morning." Ben Schreckinger, "Inside Donald Trump's Election Night War Room," *GQ,* November 7, 2017.

52 **A week later:** Smartcheckr.com was registered on November 15, 2016, at 17:04 UTC, according to lookup.icann.org.

52 **By early January 2017:** Emails from Johnson, Smartcheckr Articles of Incorporation document, and New York State business records, specifically an entity search on the New York Department of State Division of Corporations. Smartcheckr was registered as a domestic limited liability company in the state on February 10, 2017.

52 **a search app:** This description is based on company documents prepared around that time, including a project proposal that included a screenshot of the app's search results and an investor memorandum seeking $2 million in funding, both on file with author. It described one potential application as "reporting on individuals":

> In less than one second, the client receives links to the individual's social media accounts and a brief description of his or her background along with alerts if the individual is involved with:
>
> - Activities related to violence, criminal gangs or drugs
> - Radical political or religious activities
> - Terrorist-related activities
> - Prior travel to countries or regions of concern

The system can also provide analysis of individuals based on:

- Net worth
- Payment fraud
- Employment & Profession
- Health & Fitness
- Hobbies & Avocations.

It goes on from there.

52 **He was spotted:** Will Sommer, "GOP Congressmen Meet with Accused Holocaust-Denier Chuck Johnson," Daily Beast, January 17, 2019. The story cites a photo tweeted by reporter Matt Fuller earlier that day showing Johnson walking with Representatives Phil Roe (then R-TN) and Andy Harris (R-MD) and noting that the congressmen had waited for Johnson to get through security.

53 ***Forbes* reported:** Ryan Mac and Matt Drange, "A Troll Outside Trump Tower Is Helping to Pick Your Next Government," *Forbes,* January 9, 2017.

53 **"Ask Me Anything" session:** "I'm Chuck Johnson, Founder of GotNews and WeSearchr.com. Ask Me Anything!," reddit.com/r/altright, January 27, 2017. The Altright subreddit was later banned by Reddit, and the AMA is no longer available there and was instead accessed at https://archive.is/Rrawk.

53 **"alt-right or Antifa":** Author's interview with person involved with the DeploraBall who spoke on condition of not being named, 2021.

53 **"great people of all backgrounds":** "The DeploraBall," Eventbrite, https://web.archive.org/web/20161228000726/https:/www.eventbrite.com/e/the-official-deploraball-party-tickets-30173207877.

54 **"sweeping fear":** Author's interview with Lacy MacAuley, 2021.

54 **what *Atlantic* reporters had captured:** Daniel Lombroso and Yoni Appelbaum, "'Hail Trump!': White Nationalists Salute the President-Elect," *The Atlantic,* November 21, 2016.

55 **Over beers there:** Benjamin Freed, "Activist Group: Stink-Bomb Plot Was Meant to Fool James O'Keefe," *Washingtonian,* January 16, 2017.

55 **Tyler had been filming:** "Part I: Hidden-Camera Investigation Exposes Groups Plotting Violence at Trump Inauguration," Project Veritas, January 16, 2017.

55 **ultimately pled guilty:** Valerie Richardson and Andrea Noble, "Third Protester Pleads Guilty in Plot to Wreck Trump Inaugural Ball," *Washington Times,* March 8, 2017.

55 **for just thirty minutes:** Rosie Gray, "The 'New Right' and the 'Alt-Right' Party on a Fractious Night," *The Atlantic,* January 20, 2017.

56 **described as a "strongman":** Damon Linker, "What American Conservatives Really Admire About Orbán's Hungary," *The Week,* August 10, 2021.

56 **"Trump before Trump":** Gareth Browne, "Orban Is the Original Trump, Says Bannon in Budapest," The National, May 24, 2018.

56 **"Hi Gene":** Hoan Ton-That, email to Eugene Megyesy, April 3, 2017, provided to the author by Charles Johnson.

57 **billionaire George Soros:** After growing up in Hungary, Soros moved abroad and worked in banking, eventually becoming a billionaire after launching his own hedge fund. He is the subject of many conservative conspiracy theories. Emily Tamkin, "Who's Afraid of George Soros?," *Foreign Policy,* October 10, 2017.

57 **wanted it out of his country:** The next year, in March 2018, the Open Society Foundations pulled its staff out of Budapest after the government passed legislation nicknamed "Stop Soros." "The legislation, invoking national security interests, would block any organization from advising or representing asylum seekers and refugees without a government license," the organization wrote on its website. "The government has indicated that these new laws are intended to stop the work of leading Hungarian human rights organizations and their funders, including the Open Society Foundations." "The Open Society Foundations to Close International Operations in Budapest," Open Society Foundations, May 15, 2018.

57 **but then was ousted:** Kirsten Grind and Keach Hagey, "Why Did Facebook Fire a Top Executive? Hint: It Had Something to Do with Trump," *Wall Street Journal,* November 11, 2018.

57 **funding a meme-maker:** Gideon Resnick and Ben Collins, "Palmer Luckey: The Facebook Near-Billionaire Secretly Funding Trump's Meme Machine," Daily Beast, September 22, 2016.

57 **he was working on a new company:** Nick Wingfield, "Oculus Founder Plots a Comeback with a Virtual Border Wall," *New York Times,* June 4, 2017.

58 **It has long annoyed:** One of the earliest automated bots was the World Wide Web Wanderer, created in 1993 by an MIT student who was cataloging URLs to keep tabs on the size of the burgeoning internet. It "helped spark a debate about whether bots were a positive influence on the internet." Dana Mayor, "World Wide Web Wanderer Guide: History, Origin, & More," History-Computer, December 15, 2022, https://history-computer.com/world-wide-web-wanderer-guide/.

58 **a commercial service:** There are companies that help scrapers evade detection by giving their web crawlers digital cover. An infamous example is Hola, a company that offered consumers a free virtual private network (VPN) to hide their tracks on the internet. If a woman wanted to read an ex-boyfriend's blog, for example, without showing up in the visitor logs, she could use Hola's VPN to route her internet traffic through a different IP address and obscure her digital footprints. There was a trade-off: The millions of people using the "free" privacy tool were giving Hola access to their internet connections, and the company was making its money, unbeknownst to its users, by selling those connections, or "proxies," to people like Hoan Ton-That to camouflage their bots.

59 **led nowhere:** Years later, Eugene Megyesy said he did not recall Smartcheckr and couldn't remember the pitch. Email exchange with the author in 2022. "Maybe they did, but I do no recall content or any response," he wrote.

Chapter 6: The Snooper Bowl (2001)

60 **a typo-ridden press release:** GraphCo Technologies / Barry Hodge, "On January 28th, Criminals No Longer Another Face in the Tampa Stadium Crowd," Viisage Technology, January 29, 2001, Wayback Machine, https://web.archive.org/web/20030811214704/http:/www.viisage.com/january_29_2001.htm.

60 **"not everyone comes":** The press release attributes this quote to David Watkivns [*sic*], G-TEC's managing director.

60 **Viisage, a company that:** Author's interview with Tom Colatosti, 2022.

61 **The *St. Petersburg Times* broke:** Here are the articles cited in this section: Robert Trigaux, "Cameras Scanned Fans for Criminals," *St. Petersburg Times,* January 31, 2001; Bob Kappstatter, "Tampa Cops Recorded Every Fan's Face—Snooper Bowl," *Daily News* (New York), February 2, 2001.

61 **none of them were stopped:** Lisa Greene, "Face Scans Match Few Suspects," *St. Petersburg Times,* February 16, 2001.

61 **"It's just another high-tech tool":** Tampa police spokesman Joe Durkin, quoted in Louis Sahagun and Josh Meyer, "Secret Cameras Scanned Crowd at Super Bowl for Criminals," *Los Angeles Times,* February 1, 2001.

61 **"Oh my God":** Bruce Schneier, quoted in ibid.

61 **"troubled":** Erwin Chemerinsky, quoted in ibid.

62 ***The New York Times*' editorial board:** "Super Bowl Snooping," *New York Times,* February 4, 2001.

62 **"There's no such thing":** Author's interview with Tom Colatosti, 2022.

62 **a plan to use facial recognition:** Barnaby Feder, "Maker of Crowd Scanner Is on the Defensive Again," *New York Times,* February 18, 2002.

62 **The American Civil Liberties Union (ACLU) of Florida:** "ACLU Calls for Public Hearings on Tampa's 'Snooper Bowl' Video Surveillance," ACLU, February 1, 2001.

62 **"totalitarian societies":** Howard Simon, quoted in Vickie Chachere, "High-Tech Security System Makes Some Nervous of Big Brother," Associated Press, February 12, 2001.

62 **Dick Armey:** Angela Moore, "Ybor's Eyes an Invasion, U.S. House Leader Says," *St. Petersburg Times,* July 3, 2001.

62 **rare joint statement:** "Proliferation of Surveillance Devices Threatens Privacy," ACLU, July 11, 2001.

62 **Armey asked the General Accounting Office:** GAO, letter in response to Armey. Subject: Federal Funding for Selected Surveillance Technologies, U.S. Government Accountability Office, March 14, 2002, https://www.gao.gov/assets/gao-02-438r.pdf.

62 **"In the first week":** Charles Guenther, director of surveillance for Trump Marina Casino, quoted in Charles Piller, Josh Meyer, and Tom Gorman, "Criminal Faces in the Crowd Still Elude Hidden ID Cameras," *Los Angeles Times,* February 2, 2001.

62 **When the Tampa Police Department:** Edward Lewine, "Face-Scan Systems' Use Debated," *St. Petersburg Times,* December 8, 2001.

63 **One reporter found:** Greene, "Face Scans Match Few Suspects."

63 **was worried:** Author's interview with Joseph Atick, 2021.

63 **Atick kept a diary:** Excerpts from Joseph Atick's diary shared with the author.

63 **"Immediately there was":** Author's interview with Joseph Atick, 2021.

65 **security theater:** Bruce Schneier, *Beyond Fear: Thinking Sensibly About Security in an Uncertain World* (New York: Copernicus, 2003), 38.

65 **"We have had zero percent":** Barry Steinhardt, quoted in Lewine, "Face-Scan Systems' Use Debated."

65 **"While company officials":** Leonora LaPeter, "Interest in Facial Scans Surges," *St. Petersburg Times,* September 20, 2001.

65 **"What we fear":** Mike Pheneger, quoted in ibid.

66 **In 1987:** "Federal Funding for Selected Surveillance Technologies," U.S. Government Accountability Office, March 14, 2002, https://www.gao.gov/products/gao-02-438r.

66 **in November:** "Technology Assessment: Using Biometrics for Border Security," U.S. Government Accountability Office, November 15, 2002, https://www.gao.gov/products/gao-03-174.

66 **Atick's system:** By this point, Visionics had become Identix.

66 **The largest deployment:** The GAO's November 2002 report, cited above, said that Atick's system had been installed in 1998, but a book by *Washington Post* reporter Robert O'Harrow said it had been installed in 1999. It was a first-of-its-kind deployment, but by the summer of 2001, the database of deviants whose faces the authorities were trying to match numbered only sixty people, so it was more of a novelty than an aid to the surveillance unit's day-to-day work. Robert O'Harrow, Jr., *No Place to Hide* (New York: Free Press, 2005).

66 **"creepy":** Author's interview with Colatosti, 2022.

67 **A black-and-white photo:** "NIST Timeline," National Institute of Standards and Technology, https://www.nist.gov/timelinelist.

67 **So a scientist:** Author's interview with P. Jonathon Phillips, 2020.

67 **FERET:** "Face Recognition Technology (FERET)," National Institute of Standards and Technology, https://www.nist.gov/programs-projects/face-recognition-technology-feret.

68 **a seventy-one-page report:** Duane M. Blackburn, Mike Bone, and P. Jonathon Phillips, "Facial Recognition Vendor Test 2000: Evaluation Report," Defense Technical Information Center, February 16, 2001, https://apps.dtic.mil/sti/pdfs/ADA415962.pdf.

68 **more than thirty-seven thousand Mexican citizens:** P. Jonathon Phillips et al., "Face Recognition Vendor Test 2002: Evaluation Report," National Institute of Standards and Technology, March 2003, https://nvlpubs.nist.gov/nistpubs/Legacy/IR/nistir6965.pdf.

69 **"lessening the possibility":** "Privacy Impact Assessment: Integrated Biometric System," U.S. Department of State, November 2021, https://www.state.gov/wp-content/uploads/2022/02/Integrated-Biometric-System-IBS-PIA.pdf.

69 **thieves trying to obtain multiple licenses:** *Gass v. Registrar of Motor Vehicles,* Commonwealth of Massachusetts Appeals Court, January 7, 2013.

70 **"We didn't call it bias":** Author's interview with Patrick Grother, 2021. The reports were focused on how facial recognition worked generally. To see how it was performing on specific groups of people, a reader would have had to closely examine the tail end of the lengthy Facial Recognition Vendor reports. There, the differential performance was meticulously plotted on graphs of each algorithm's performance across an array of categories, including gender and age, but it wasn't spelled out for a lay reader.

70 **After creating:** *Combined DNA Index System: Operational and Laboratory Vulnerabilities,* Audit Report 06-32, U.S. Department of Justice, Office of the Inspector General, Audit Division, May 2006, https://oig.justice.gov/reports/FBI/a0632/final.pdf; "Status of IDENT/IAFIS Integration," U.S. Department of Justice, December 7, 2001, https://oig.justice.gov/reports/plus/e0203/back.htm.

70 **to identify perpetrators:** By 2005, law enforcement agencies had used the fingerprint database more than 100 million times and the DNA database had produced hits in nearly thirty thousand investigations. "Fingerprint System Developed by Lockheed Martin Helps FBI Reach Milestone of 100 Million Print Searches," Lockheed Martin, August 22, 2005; *Combined DNA Index System Operational and Laboratory Vulnerabilities,* Audit Report 06-32, U.S. Department of Justice, Office of the Inspector General, May 2006.

70 **"Blacks are easier":** Patrick J. Grother, George W. Quinn, and P. Jonathon Phillips, "Multiple-Biometric Evaluation (MBE) 2010: Report on the Evaluation of 2D Still-Image Face Recognition Algorithms," National Institute of Standards and Technology, August 24, 2011, https://nvlpubs.nist.gov/nistpubs/Legacy/IR/nistir7709.pdf, 56.

70 **"Psychological research indicates":** P. Jonathon Phillips et al., "An Other-Race Effect for Face Recognition Algorithms," *ACM Transactions on Applied Perception* (2009), https://www.nist.gov/publications/other-race-effect-face-recognition-algorithms.

71 **Technical documents warned:** "What Is IPSFRP?," FBI, obtained by the Electronic Frontier Foundation and available at "FBI NGI Description of Face Recognition Program," https://www.eff.org/document/fbi-face-recognition-documents.

71 **There were success stories:** Kelly Bauer, "Facial Recognition Software Used to Bust Rush Street Bank Robber, Feds Say," DNA Info, June 10, 2016; Sun-Times Wire, "Serial Chicago Bank Robber Gets 108 Months in Prison for String of Heists," *Chicago Sun-Times,* February 15, 2018; Ryan Lucas, "How a Tip—and Facial Recognition Technology—Helped the FBI Catch a Killer," NPR, August 21, 2019.

71 **no real-world audit:** The argument that the performance of facial recognition algorithms must be tested in the real world, not just in labs, has been particularly well made by Inioluwa Deborah Raji, in writings, talks, and interview with the author, 2022.

Chapter 7: The Supercomputer Under the Bed

72 **"the tiniest room of all time":** Author's interview with Hoan Ton-That, 2021.

72 **a new baby:** Author's interview with Charles Johnson, 2021.

73 **"It's going to sound":** Author's interview with Hoan Ton-That, 2021.

73 **Ever since then:** Cade Metz, *Genius Makers: The Mavericks Who Brought AI to Google, Facebook, and the World* (New York: Dutton, 2021).

74 **"We intend to maintain OpenFace":** Brandon Amos, Bartosz Ludwiczuk, and Mahadev Satyanarayanan, "OpenFace: A General-Purpose Face Recognition Library with Mobile Applications," Carnegie Mellon University, June 2016, https://elijah.cs.cmu.edu/DOCS/CMU-CS-16-118.pdf.

74 **"Please use responsibly!":** When told about that, Heather Douglas, a philosopher of science ethics, burst out laughing. "That's great if you have a contract and a lawyer that you are willing to hire to track people down," she said. Even then, "How are you going to sue China?" In her book *Science, Policy, and the Value-Free Ideal,* she documented scientists' failure to reckon with the consequences of their work outside a laboratory. From the physicists in the Manhattan Project who built the atom bombs that the U.S. government dropped on Hiroshima and Nagasaki to the doctors who let syphilis go untreated to see its full effects on Black patients at the Tuskegee Institute, scientific researchers in pursuit of the unknown have historically privileged knowledge at all costs. Over the last two decades, Douglas has seen accountability spring up in physics, medicine, and biology, with guardrails put up, for example, around genetically modifying human embryos. But the reckoning hasn't fully come for computer scientists yet.

"OpenFace," GitHub, https://cmusatyalab.github.io/openface/; author's interview with Douglas, 2022.

74 **"I couldn't have figured":** Author's interview with Hoan Ton-That, 2021.

75 **"overlapping tech circles":** Author's email exchanges with Terence Z. Liu, October 2021.

75 **"a person doing one calculation":** Steve Lohr, "Move Over, China: U.S. Is Again Home to World's Speediest Supercomputer," *New York Times,* June 8, 2018.

75 **"I needed to put":** Author's email exchanges with Terence Z. Liu, October 2021.

76 **"stole a lot of computer time":** "He also, funny note, stole a lot of computer time on their supercomputer where he graduated from. So that helped early on with building the neural network." Author's interview with Hoan Ton-That, 2020.

76 **"the supercomputer under my bed":** Author's email exchanges with Terence Z. Liu, October 2021.

76 **never heard of it:** "I am not aware of any supercomputer he built while here." Author's email exchanges with Sanjay Khare, professor in the department of physics and astronomy at the University of Toledo, 2022.

76 **Liu was much better:** "Terence is really quick on mathematical stuff," Ton-That said. "He will take what I find and have a better idea of how to get it working, tune the parameters, figure out the right loss functions, and test it." Author's interview with Hoan Ton-That, 2021.

76 **researchers at Microsoft:** "MS-Celeb-1M: Challenge of Recognizing One Million Celebrities in the Real World," Microsoft, June 29, 2016. See also Yandong Guo et al., "Ms-celeb-1m: A Dataset and Benchmark for Large-Scale Face Recognition," Computer Vision–ECCV 2016, 14th European Conference, Amsterdam, Netherlands, October 11–14, 2016, Proceedings, part 3, 14, Springer International Publishing, 2016.

76 **there were also journalists and activists:** Adam Harvey and Jules LaPlace, "MS-CELEB-1M," Exposing.ai, 2021.

76 **"This is crazy":** Author's interview with Hoan Ton-That, 2021.

76 **"Out of 10 million faces":** Wayback Machine, https://web.archive.org/web/20170623075224/http://terencezl.ngrok.io/facesearch/.

77 **"Check out this cool app":** Davis King, @nulhom, Twitter, June 15, 2017, https://twitter.com/nulhom/status/875427341538910208.

77 **grew quickly:** Emma Roller, "The Spawn of Facebook and PayPal," Slate, August 5, 2013; Eric Levenson, "Why the Venmo Newsfeed Is the Best Social Network Nobody's Talking About," *The Atlantic,* April 29, 2014.

78 **privacy proponents were alarmed:** Concerned technologists kept coming up with stunts to help make people aware of their exposure on Venmo. Two coders created a site called Vicemo—a collection of public transactions they had scraped from Venmo that mentioned "drugs, booze, and sex." Another coder collected more than 200 million public transactions in 2017 for an online project called Public by Default, in which she spotlighted an extended back-and-forth between a couple having a nasty breakup and a drug dealer's explicit Venmo requests to his clients.

78 **how the company may exploit it:** A professor named Joseph Turow was the first person to study this. When he polled home internet users about privacy policies back in 2003, 57 percent believed that a website with a privacy policy wouldn't share their personal info with third parties. When Pew Research Center asked Americans a variation of the same question a decade later, 52 percent of them again said, incorrectly, that a privacy policy means that a company will keep personal information confidential. Sadly, this is not the case. "Half of Americans don't know what a privacy policy is," Pew declared in 2014. Joseph Turow, "Americans Online Privacy: The System Is Broken," Annenberg Public Policy Center, University of Pennsylvania, June 2003; Aaron Smith, "Half of Online Americans Don't Know What a Privacy Policy Is," Pew Research Center, December 4, 2014.

78 **"It's annoying":** Author's interview with Hoan Ton-That, 2021.

78 **"You could hit this one URL":** Ibid.

79 **remained that way for years:** Venmo took serious steps to protect users' profiles only in 2021, after reporters at BuzzFeed News found the accounts of President Joe Biden and his wife, Jill Biden. The reporters showed that it was

possible to create an extensive map of the extended Biden family, their friends, and their associates—a national security risk—because their friend lists were exposed. After that, Venmo announced that it would make its users' contacts private and no longer make any transactions public to the world. It took the Secret Service calling for the company to take privacy concerns seriously.

79 **"Scraped 2.1 million faces":** Hoan Ton-That, email to Charles Johnson and Richard Schwartz, June 8, 2017. Provided by Johnson, on file with author.

79 **"I got a prototype working":** Hoan Ton-That, email to Palmer Luckey and Charles Johnson, June 13, 2017. Provided by Johnson, on file with author.

79 **"It means we can find":** Hoan Ton-That, email to Peter Thiel and Charles Johnson, June 14, 2017. Provided by Johnson, on file with author.

80 **one of Thiel's lieutenants emailed:** "Peter asked me to reach out about potentially investing $200k in Smartchekr's [*sic*] financing round. Do you have a deck, term sheet and/or investment documents for me to review?" Bryce Steeg, email to Hoan Ton-That, July 20, 2017, forwarded to Charles Johnson the same day. Provided by Johnson, on file with author.

80 **Over lunch at a Greek restaurant:** Author's interview with Sam Waxman, 2022.

80 **called Smartcheckr Corp, Inc.:** Certificate of Incorporation of Smartcheckr Corp, Inc., State of Delaware Secretary of State Division of Corporations, August 3, 2017.

80 **Johnson had no idea:** Author's interviews with Charles Johnson, 2021–2022.

80 **"officially closed money":** Hoan Ton-That, email to Palmer Luckey and Charles Johnson, August 14, 2017. Provided by Johnson, on file with author.

81 **nearly a year after:** Clearview AI Investor Deck, July 2018. On file with author.

81 **"I had to swap":** Author's interview with Hoan Ton-That, 2020.

Chapter 8: The Only Guy Who Saw It Coming (2006–2008)

82 **"No, thank you":** Author's interview with James Ferg-Cadima, 2021. Much of this chapter comes from Ferg-Cadima's recollection of this time and was corroborated by Adam Schwartz, who worked with Ferg-Cadima.

83 **"There's a handful":** Author's interview with James Ferg-Cadima, 2021.

83 **prohibit employers and insurance companies:** It was passed in 2008. "The Genetic Information Nondiscrimination Act of 2008," U.S. Equal Employment Opportunity Commission.

83 **Texas Capture or Use:** It was passed in 2007 and went into effect in 2009. "80(R) HB 2278—Enrolled Version," https://capitol.texas.gov/tlodocs/80R/billtext/html/HB02278F.HTM.

84 **"If that were compromised":** Author's interview with James Ferg-Cadima, 2021.

84 **Pay By Touch:** "Pay By Touch," Crunchbase; Becky Bergman, "Pay By Touch Sees Biometrics as Key to Secure Transactions," *Silicon Valley Business*

Journal, June 19, 2005; Ben Worthen, "Another Hot New Technology Turns Cold," *Wall Street Journal,* March 19, 2008.

84 **Ferg-Cadima was troubled:** Author's interview with James Ferg-Cadima, 2021.

84 **had a lengthy court history:** "Five creditors sued Rogers, but some had trouble collecting, records show. . . . Rogers also had conflicts with women. In 1997, a girlfriend accused him of slamming her head against the car window during an argument in Rogers' BMW sedan, according to an arrest report cited in the lawsuits. He pleaded guilty to disorderly conduct." Lance Williams, "How 'Visionary' Raised—and Lost—a Fortune," *San Francisco Chronicle,* December 7, 2008.

84 **shocking news**: Matt Marshall, "Pay By Touch in Trouble, Founder Filing for Bankruptcy," VentureBeat, November 12, 2007. Brad Meikle, "Pay By Touch Faces Bankruptcy Petition," *Buyouts,* November 19, 2007.

84 **"Then all my red flags":** Author's interview with James Ferg-Cadima, 2021.

84 **made off with:** Andrea Peterson, "OPM Says 5.6 Million Fingerprints Stolen in Cyberattack, Five Times as Many as Previously Thought," *Washington Post,* September 23, 2015.

84 **"All the stars":** Author's interview with James Ferg-Cadima, 2021.

85 **"Founder and chairman John Rogers":** Nicholas Carlson, "Pay By Touch to Close Up Shop?," Gawker, November 9, 2007.

85 **Rogers had hired his mother:** Author's interview with Ajay Amlani, 2021. Amlani worked for Pay By Touch during that time period.

85 **"constant abuse of drugs":** *KOK Sheridan Way Investments, LLC v. Solidus Networks, Inc. DBA Pay-by-Touch,* Superior Court of California, June 30, 2009.

85 **The company also had:** Author's interview with Ajay Amlani, 2021.

85 **Some Christians demurred:** Ibid. See also Hutch Carpenter, "Farewell, Pay By Touch, Farewell," I'm Not Actually a Geek, March 19, 2008.

85 **Ferg-Cadima saw opportunity:** Author's interview with James Ferg-Cadima, 2021.

85 **"a retina or iris scan":** 740 ILCS 14 / Biometric Information Privacy Act, Illinois Compiled Statutes, Illinois General Assembly, https://www.ilga.gov/legislation/ilcs/ilcs3.asp?ActID=3004&ChapterID=57.

85 **"We wanted to build":** Author's interview with James Ferg-Cadima, 2021.

86 **worked the halls:** Author's interview with James Ferg-Cadima, 2021.

86 **The bill was introduced:** It passed both houses on July 10, 2008. *Legislative Synopsis and Digest of the Ninety-fifth General Assembly,* https://www.ilga.gov/ftp//Digest/95thFinalDigest.pdf.

86 **Only Illinois's had teeth in it:** Author's interview with Albert Gidari, 2021.

86 **The company that had been known:** "Pay By Touch Fades into History as Lenders Buy Core Assets," Digital Transactions, April 7, 2008.

86 **changed its focus:** Author's interview with Ajay Amlani, 2021.

86 **the company was eventually acquired:** "Kroger Announces Acquisition of

YOU Technology," Cision, February 11, 2014. "Inmar Rebrands to Inmar Intelligence," Inmar Intelligence, January 27, 2020.

87 **"To this day":** Author's interview with James Ferg-Cadima, 2021.

Chapter 9: Death to Smartcheckr

88 **Holly Lynch was thinking:** Author's interviews with Holly Lynch, 2020–2021.

89 **A few of Schwartz's friends:** Author's interviews with Hoan Ton-That, Richard Schwartz, and Charles Johnson. The narrative is also based on a company document prepared for potential investors, on file with author.

89 **"I got Giuliani elected":** Holly Lynch's recollection of the conversation during interview with the author, 2020.

89 **"I had just come out":** Author's interview with Holly Lynch, 2021.

89 **"the year of the woman":** That was true. In the 2018 midterm elections, a record-breaking number of women were running for office. Another political newcomer, a twenty-nine-year-old former bartender named Alexandria Ocasio-Cortez who leaned socialist, was running in the congressional district next door to Lynch's. Ocasio-Cortez, nicknamed AOC, had virtually no funding, but she went on to beat the third most powerful Democrat in Congress, who, like Nadler, had held his seat for twenty years.

89 **After Schwartz told Lynch:** Author's interview with Holly Lynch, 2021.

89 **$5,000 a month:** Veritas Strategic Partners invoice, October 2017; Veritas Strategic Partners c/o Richard Schwartz invoice, November 2017. On file with author.

90 **Lynch was discomforted:** Author's interview with Holly Lynch, 2021.

90 **a brochure for "Smartcheckr Corp.":** Richard Schwartz, email to Holly Lynch, October 21, 2017. Provided by Lynch, on file with author.

91 **at a cost of $24,000:** Infomentum LLC invoice, addressed to Richard Schwartz at Veritas Strategic Partners, October 30, 2017. On file with author.

91 **"path to victory":** Email with poll analysis memo sent from Infomentum to Hoan Ton-That, Richard Schwartz, and Douglass Mackey, November 9, 2017. On file with the author.

92 **"Is that legal?":** Holly Lynch's recollection of the conversation during interview with the author, 2021.

92 **It was a standard offering:** Author's interview with Daniel Kreiss, a professor at the University of North Carolina at Chapel Hill who studies the use of data by political campaigns, 2021.

92 **Cambridge Analytica claimed:** Alex Hern, "Cambridge Analytica: How Did It Turn Clicks into Votes?," *The Guardian,* May 6, 2018; Brian Resnick, "Cambridge Analytica's 'Psychographic Microtargeting': What's Bullshit and What's Legit," Vox, March 26, 2018; Timothy Revell, "How Facebook Let a Friend Pass My Data to Cambridge Analytica," *New Scientist,* April 16, 2018.

92 **on whose campaign:** Issie Lapowsky, "What Did Cambridge Analytica Really Do for Trump's Campaign?," *Wired,* October 26, 2017.

93 **"I'm very concerned":** Holly Lynch's recollection in interview with the author, 2021.

93 **feeling like an inexperienced candidate:** Author's interview with Holly Lynch, 2021.

93 **described himself:** David Weigel, "Ryan's 'Pro-White' Primary Foe Denounced by Breitbart After His Anti-Semitic Tweets," *Washington Post,* December 27, 2017.

93 **The brochure:** "Smartcheckr Consulting Group," on file with author and available on Wayback Machine, https://web.archive.org/web/20190423184841/https://christophercantwell.com/wp-content/uploads/2018/04/Smartcheckr-Paul-Nehlen-Oct-19.pdf.

93 **"magic beans":** Christopher Cantwell's *The Radical Agenda,* April 3, 2018, since removed from the web. Paul Nehlen talking about Douglass Mackey: "He reached out to me through somebody else and said I'd like to do some consulting work with you. It'll basically cost about $2,500 bucks a month and I looked at it, he pitched it to me and I thought, 'This guy's trying to sell me magic beans. I don't agree with it. I don't think it's going to work.' . . . But I figured I'm going to give the guy a shot and I gave him full access to my Facebook through business manager for three months. . . . He did nothing, he did absolutely nothing." Paul Nehlen's campaign spending report, via the FEC, did not reflect any payments to Mackey, Veritas Strategic Partners, or Smartcheckr.

93 **nothing to show for it:** When the company was asked about the Nehlen incident years later, its representative said that Mackey had been a consultant to the company for a short time, that the company hadn't known of his alter ego, and that he had sent an "unauthorized proposal" to Nehlen. Ton-That said that "the technology described in the proposal did not even exist." (So Holly Lynch's sense that she had been exploited may not have been off base.)

93 **Mackey had an alter ego:** Luke O'Brien, "Trump's Most Influential White Nationalist Troll Is a Middlebury Grad Who Lives in Manhattan," HuffPost, April 5, 2018.

94 **outlandish claims about Hillary Clinton:** Years later, in January 2021, Mackey, aka Vaughn, was charged with election interference, a federal offense, for tweeting that voters for Hillary Clinton could avoid long lines by voting by text or social media and providing the means to do so. That, of course, is not a legal way to vote. At least 4,900 people actually texted the number he provided with Clinton's name, according to federal prosecutors. "Social Media Influencer Charged with Election Interference Stemming from Voter Disinformation Campaign," U.S. Department of Justice, January 27, 2021, https://www.justice.gov/opa/pr/social-media-influencer-charged-election-interference-stemming-voter-disinformation-campaign.

94 **"Right Wing Gossip Squad":** Zanting, "Rudy Giuliani Ex-Advisor Richard Schwartz Worked With the Alt-Right," AnimeRight News, May 7, 2018,

https://web.archive.org/web/20190330154815/https://animeright.news/zanting/rudy-giuliani-ex-advisor-richard-schwartz-worked-with-the-alt-right/.

94 **All of the bad publicity:** A top hit on Google for "Smartcheckr" was Eyes on the Right, "The Alt-Right Continues to Disintegrate After White Nationalist Troll 'Ricky Vaughn' Gets Unmasked," Angry White Men, April 6, 2018, https://angrywhitemen.org/2018/04/06/the-alt-right-continues-to-disintegrate-after-white-nationalist-troll-ricky-vaughn-gets-unmasked/. Commenters on 4Chan discussed it as well: https://archive.4plebs.org/pol/thread/169758083/.

94 **evidently spooked:** Ton-That and Schwartz declined to comment about this episode in their company's history. Ton-That said they changed the name from Smartcheckr to Clearview AI only because it was "better branding." Author's interview with Hoan Ton-That for *New York Times,* 2020.

94 **A search of his name:** Author's Google search, 2019.

94 **a different Richard Schwartz:** Sapna Maheshwari, "Richard J. Schwartz, Who Expanded Jonathan Logan's Dress Empire, Dies at 77," *New York Times,* October 5, 2016.

94 **registered a new website address:** WHOIS shows that Clearview.ai was registered on December 16, 2017.

94 **"Are you trying to Eduardo Saverin me?":** As recollected by Charles Johnson in an interview with the author, 2021.

95 **Ton-That shaken and tearful:** Author's interview with Richard Schwartz, 2021.

95 **Johnson had a lot going on:** Author's interview with Charles Johnson, 2021.

95 **news outlets expressed shock:** The Florida lawmaker was Matt Gaetz. "GOP Lawmaker Condemned for Inviting Holocaust Denier to State of the Union," *The Guardian,* February 1, 2018.

95 **a flurry of introductions:** Emails provided by Charles Johnson to author.

95 **"No Dorsey for now":** Hoan Ton-That, email to Charles Johnson, June 23, 2018. Provided by Johnson, on file with author.

95 **they called their lawyer:** Author's interview with Sam Waxman, 2022.

96 **"Let's finish this off":** Hoan Ton-That, email to Charles Johnson, November 25, 2018. Provided by Johnson, on file with author.

96 **By the end of 2018:** Author's interview with Hoan Ton-That, 2021.

PART II: TECHNICAL SWEETNESS

Chapter 10: The Line Google Wouldn't Cross (2009–2011)

99 **"Until now":** Google, "Google Goggles," YouTube, December 7, 2009, https://www.youtube.com/watch?v=Hhgfz0zPmH4.

99 **but when people actually started:** David Pierce, "How Google Goggles Won, Then Lost, the Camera-First Future," *Wired,* October 27, 2017.

100 **a tech journalist from CNN:** Mark Milian, "Google Making App That Would Identify People's Faces," CNN, March 31, 2011.

100 **"People are asking":** Hartmut Neven, quoted in ibid.

100 **Years earlier:** "Celebrating 15 Years of Street View," Google, https://www.google.com/streetview/anniversary/.

100 **cutting-edge feature:** Siva Vaidhyanathan, *The Googlization of Everything (And Why We Should Worry),* updated ed. (Berkeley: University of California Press, 2012), 98.

100 **The Borings:** Opinion, *Boring v. Google Inc.,* United States Court of Appeals, Third Circuit, January 28, 2010.

101 **a blurring option:** "Street View: Exploring Europe's Streets with New Privacy Safeguards," Google Europe Blog, June 12, 2009, https://publicpolicy.googleblog.com/2009/06/street-view-exploring-europes-streets.html.

101 **Hundreds of thousands:** "How Many German Households Have Opted-Out of Street View?," Google Europe Blog, October 21, 2010, https://europe.googleblog.com/2010/10/how-many-german-households-have-opted.html.

101 **"Google's cool":** "Street View," Deutsche Welle, November 23, 2010, https://www.dw.com/en/wayward-google-fans-pelt-houses-with-eggs/a-6258069.

101 **according to Neven:** Milian, "Google Making App That Would Identify People's Faces."

101 **"Picasa's facial recognition":** "Picasa Refresh Brings Facial Recognition," TechCrunch, September 2, 2008.

102 **When other journalists reached out:** Vlad Savov, "Google Working on a Face Recognition App That Leads to Your Personal Info? (Update: Google Says 'No')," Engadget, March 31, 2011; Hayley Tsukayama, "Google Denies CNN Report of Facial Recognition App," *Washington Post,* March 31, 2011.

102 **"We built that technology":** "D9 Video: Eric Schmidt on Privacy," *Wall Street Journal,* June 1, 2011. See also Bianca Bosker, "Facial Recognition: The One Technology Google Is Holding Back," Huffington Post, June 1, 2011.

102 **When Google launched:** Jason Beahm, "Google Buzz Lawsuit Settles," FindLaw, November 4, 2010.

102 **secretly sucking up data:** The Street View cars' data collection, dubbed the Wi-Spy scandal by the press, led to numerous fines for the company even though it claimed to have gathered the information inadvertently. Notice of Apparent Liability for Forfeiture, in the Matter of Google Inc., before the Federal Communications Commission, April 13, 2012.

103 **"It's simple to find":** Author's interview with Henry Schneiderman, 2020.

103 **One of Schneiderman's fellow grad students:** Henry Rowley, Shumeet Baluja, and Takeo Kanade, "Neural Network-Based Face Detection," *IEEE Transactions on Pattern Analysis and Machine Intelligence* 20, no. 1 (1998): 23–38.

104 **his thesis paper:** Henry Schneiderman, "A Statistical Approach to 3D Object Detection Applied to Faces and Cars," PhD thesis, Carnegie Mellon University, May 10, 2000.

105 **digital video made it easier:** Author's interview with Henry Schneiderman, 2020.

105 **But he came to regret:** Ibid.

105 **Law enforcement agencies that had licensed:** "A leading face algorithm,

Pittsburgh Pattern Recognition (PittPatt), developed in 2004 by Carnegie Mellon University with federal funds, was tweaked differently by intelligence and law enforcement agencies to suit their varying needs." "History of NIJ Support for Face Recognition Technology," National Institute of Justice, 2020, https://nij.ojp.gov/topics/articles/history-nij-support-face-recognition-technology.

106 **"star performer":** Author's interview with former IARPA scientist who spoke on condition of not being named, 2021.

106 **survival was important:** Years later, in 2013, when two homemade pressure cooker bombs exploded at the Boston Marathon, killing three people and injuring hundreds more, a chaotic manhunt ensued for those responsible, brothers aged twenty-six and nineteen, who had emigrated to the country a decade earlier. Schneiderman was told that PittPatt had played a role in the hunt.

106 **Acquisti was fascinated:** Author's interview with Alessandro Acquisti, 2021.

106 **the "privacy paradox":** The term had been coined in 2001 by a researcher at Hewlett-Packard who interviewed people then new to the internet, who expressed concern about having their online surfing tracked by companies. The researcher then asked them whether they had loyalty cards that let stores track their shopping, and they said they did, because they appreciated the discounts they got with them. Barry Brown, "Studying the Internet Experience," HP Laboratories Bristol, March 26, 2001.

106 **"On the internet":** Peter Steiner, "On the Internet, Nobody Knows You're a Dog," *The New Yorker,* July 5, 1993.

106 **uploading 2.5 billion photos:** Eric Eldon, "New Facebook Statistics Show Big Increase in Content Sharing, Local Business Pages," *Adweek,* February 15, 2010.

106 **"most appalling spying machine":** "WikiLeaks Revelations Only Tip of Iceberg—Assange," RT, May 2, 2011.

106 **Marketers loved Facebook:** Anthony Ha, "Author Clara Shih on the Facebook Era, and What It Means for Businesses," VentureBeat, March 29, 2009.

106 **"one of the holy grails":** Stephanie Rosenbloom, "On Facebook, Scholars Link Up with Data," *New York Times,* December 17, 2007.

106 **More than 90 percent:** Ralph Gross and Alessandro Acquisti, "Information Revelation and Privacy in Online Social Networks," Proceedings of the 2005 ACM Workshop on Privacy in the Electronic Society, 2005.

107 **He doubted:** Author's interview with Alessandro Acquisti, 2021.

107 **More than thirty of the ninety-three:** Alessandro Acquisti, Ralph Gross, and Fred Stutzman, "Faces of Facebook: Privacy in the Age of Augmented Reality," Black Hat presentation, 2012, https://www.blackhat.com/docs/webcast/acquisti-face-BH-Webinar-2012-out.pdf.

108 **Acquisti and his colleagues were astonished:** Author's interview with Alessandro Acquisti, 2021.

108 **"democratization of surveillance":** Alessandro Acquisti, quoted in Kashmir Hill, "Hello, Stranger," *Forbes,* August 24, 2011.

108 **It had been acquired:** "We are happy to announce that Pittsburgh Pattern Recognition has been acquired by Google!" PittPatt via Wayback Machine, https://web.archive.org/web/20110824231552/https:/www.pittpatt.com/. See also James Niccolai, "Google Buys Facial Recognition Company PittPatt," CSO Online, July 22, 2011.

109 **It was only after:** Author's interview with Henry Schneiderman, 2021.

109 **"When is it fair":** Schneiderman's recollection of his thought process at the time in an interview with the author, 2021.

109 **Google warned its own employees:** Author's interview with former Google employee on the company's privacy team who asked not to be named, 2020.

109 **"It didn't take off":** Ibid.

109 **one patent application:** "Audience Attendance Monitoring Through Facial Recognition," Google patent, applied for August 7, 2014, granted November 20, 2018, https://patents.google.com/patent/US10134048B2. See also "Automatic Transition of Content Based on Facial Recognition," Google patent, applied for September 4, 2012, granted February 24, 2014, https://patents.google.com/patent/US8965170B1.

109 **Google also imagined:** "Monitoring and Reporting Household Activities in the Smart Home According to a Household Policy," Google patent, applied for March 5, 2015, granted January 16, 2018, https://patents.google.com/patent/US20160261932A1.

110 **"It's much easier":** Alessandro Acquisti during a presentation at the Federal Trade Commission, December 8, 2011, https://www.ftc.gov/sites/default/files/documents/videos/face-facts-session-3/4/120811_ftc_sess3.pdf.

Chapter 11: Finding Mr. Right

111 **David Scalzo came across:** Author's interview with David Scalzo, 2021.

111 **They gave Scalzo the pitch:** The description of the pitch is based on ibid., as well as documents the company had prepared for potential investors that were being circulated around this time. Provided by people who were pitched by the company, on file with author.

112 **"while simultaneously assessing them":** Clearview AI investor deck, 2018. Provided by a person who was pitched by the company, on file with author.

112 **"I have plenty":** Ibid.

112 **It could tell sisters apart:** Most of these examples come from a company document distributed around that time titled "Technical Overview and Examples," 2018. Provided by a person who was pitched by the company, on file with author.

112 **It found two photos:** Author's interview with David Scalzo, 2021.

112 **"very misfitty":** Author's interview with John Borthwick for *The New York Times,* 2020.

113 **"You freaked the hell":** Author's interview with Hoan Ton-That, 2021.

113 **"It's too much":** Ibid.

113 **at poker tables:** Author's interview with Charles Johnson, 2021.

113 **"The future is already here":** William Gibson, interviewed on "The Science of Science Fiction," NPR, November 30, 1999, via The Quote Investigator, January 24, 2012.

113 **hundreds of searches:** Ryan Mac, Caroline Haskins, and Logan McDonald, "Secret Users of Clearview AI's Facial Recognition Dragnet Included a Former Trump Staffer, a Troll, and Conservative Think Tanks," BuzzFeed News, March 25, 2020.

113 **Leone texted Ton-That:** An anonymous source provided screenshots of the texts between Doug Leone and Hoan Ton-That. On file with author.

114 **His name was Joe:** Joe Montana, email to Hoan Ton-That, February 25, 2019. Provided by an anonymous source, on file with author.

114 **"That's Joe Montana":** Author's interview with Hoan Ton-That, 2021.

114 **tested out Clearview:** Author's interview with John Catsimatidis for *The New York Times,* 2020.

114 **"People were stealing":** Catsimatidis, quoted in Kashmir Hill, "Before Clearview AI Became a Police Tool, It Was a Secret Plaything of the Rich," *New York Times,* March 5, 2020. The ice cream bandits had also been reported on in Reuven Fenton and Linda Massarella, "Ice Cream Bandits Are Wreaking Havoc on NYC Supermarkets," *New York Post,* August 12, 2016.

114 **"I wanted to make sure":** Hill, "Before Clearview AI Became a Police Tool, It Was a Secret Plaything of the Rich."

114 **"hilarious":** Author's interview with Andrea Catsimatidis for *The New York Times,* 2020.

114 **Catsimatidis had twenty-eight thousand contacts:** Author's interview with John Catsimatidis for *The New York Times,* 2020.

114 **Ton-That and Schwartz asked people:** Author's interview with Hoan Ton-That, 2021.

115 **"I have an app":** First We Feast, "Ashton Kutcher Gets an Endorphin Rush While Eating Spicy Wings | Hot Ones," YouTube, September 26, 2019, https://www.youtube.com/watch?v=nNhYqLbsAGk.

115 **Ton-That was annoyed:** Author's interview with Hoan Ton-That, 2021.

115 **"You almost blew":** Hoan Ton-That's recollection of the conversation in ibid.

115 **a handful of times:** Author's email exchange with Ashton Kutcher, March 2, 2022.

115 **He was put off:** Ibid. "Existing investors are integrity signals," wrote Kutcher in an email to author.

115 **One investor told an acquaintance:** Author's interview with a source who asked not to be named, 2020.

115 **Kendall Jenner liked his app:** Author's interview with Hoan Ton-That, 2021.

116 **"Whoa, dude":** Ibid.

116 **Left-leaning investors:** Author's interview with a source who asked not to be named, 2020.

116 **Investors who did due diligence:** Ibid.

116 **"In my mind":** Author's interview with Hoan Ton-That, 2021.

116 **Scalzo experimented:** Author's interview with David Scalzo for *The New York Times,* 2020.

117 **"Information is power":** Author's email exchange with David Scalzo, March 3, 2021.

117 **moms of America:** Author's interview with David Scalzo, 2023.

117 **Scalzo knew the company:** Author's interview with David Scalzo, 2021.

117 **a high-profile lawsuit:** *hiQ Labs, Inc. v. LinkedIn Corp.,* 938 F.3d 985, Ninth Circuit 2019; Drake Bennett, "The Brutal Fight to Mine Your Data and Sell It to Your Boss," Bloomberg, November 15, 2017; Jamie Williams and Amul Kalia, "Judge Cracks Down on LinkedIn's Shameful Abuse of Computer Break-in Law," Electronic Frontier Foundation, August 29, 2017; "LinkedIn Corp. v. hiQ Labs, Inc.," Electronic Privacy Information Center, https://epic.org/documents/linkedin-corp-v-hiq-labs-inc/.

118 **business advantage:** Elizabeth Pollman and Jordan M. Barry, "Regulatory Entrepreneurship," *Southern California Law Review* 90 (March 7, 2016): 383.

118 **a random collective:** Company documents prepared for potential investors. On file with author.

118 **"You can't":** Author's interview with Hal Lambert for *The New York Times Magazine,* 2020.

118 **Ton-That introduced the red-headed Johnson:** Author's interview with David Scalzo, 2021.

119 **"wicked smart":** Ibid.

119 **"they want to win":** David Scalzo's recollection of Johnson's statement in ibid.

119 **"A little bit of an evil genius":** Ibid.

119 **Peter Thiel was co-hosting:** Charlie Nash, "PICTURES: Peter Thiel Holds Fundraiser for Kris Kobach at NYC Apartment (Featuring Ann Coulter)," Mediaite, September 19, 2019.

119 **"I'm invested in that?":** David Scalzo's recollection of the conversation in interview with the author, 2021.

Chapter 12: The Watchdog Barks (2011–2012)

121 **On a wintry Thursday morning:** The author attended this workshop in 2011. Video and transcripts from the day are also archived online: "Face Facts: A Forum on Facial Recognition Technology," Federal Trade Commission, December 8, 2011, https://www.ftc.gov/news-events/events/2011/12/face-facts-forum-facial-recognition-technology. Descriptions and quotes from the event are based on the archived video.

121 **a couple of carafes of coffee:** Author's interview with Maneesha Mithal, 2022.

122 **SceneTap, which had installed:** Kashmir Hill, "Using Facial Recognition Technology to Choose Which Bar to Go To," *Forbes,* June 28, 2011.

123 **Their estimate?:** Aleecia M. McDonald and Lorrie Faith Cranor, "The Cost of Reading Privacy Policies," *I/S: A Journal of Law and Policy for the Information Society* 4, no. 3 (2008): 543–68.

123 **"We didn't tell":** Inspired by a Gamestation April Fool's joke: Bianca Bosker, "7,500 Online Shoppers Accidentally Sold Their Souls to Gamestation," Huffington Post, April 17, 2010.

123 **underresourced sheriff:** Peter Maass, "Your FTC Privacy Watchdogs: Low-Tech, Defensive, Toothless," *Wired,* June 28, 2012.

126 **"We kept screaming":** Author's interview with Maneesha Mithal, 2022.

126 **a thirty-page report:** "Facing Facts: Best Practices for Common Uses of Facial Recognition Technologies," Federal Trade Commission, October 2012, https://www.ftc.gov/sites/default/files/documents/reports/facing-facts-best-practices-common-uses-facial-recognition-technologies/121022facialtechrpt.pdf.

127 **"A mobile app that could":** Ibid., 13.

Chapter 13: Going Viral

128 **"You know what you should":** Conversation as recalled by Hoan Ton-That in interview with the author, 2020; identity of officer from NYPD documents provided in a reply to an information request to Rachel Richards, including communications sent from October 2018 to February 2020; "NYPD Clearview AI Communications," MuckRock, March 31, 2021.

128 **"Somewhere around":** Author's interview with Hoan Ton-That, 2020.

128 **Greg Besson had been:** Greg Besson, résumé on LinkedIn, https://www.linkedin.com/in/greg-besson-b03208241/.

129 **Gilbert III, was considering investing:** Investor documents that Clearview was circulating at the time. On file with author.

129 **was initially reluctant:** Author's interview with Hoan Ton-That, 2020.

129 **Besson added:** All emails quoted, reproduced, or described in this chapter, unless otherwise noted, are from the NYPD public records request, MuckRock.

129 **"crime-fighting kings":** Willard M. Oliver, *Policing America: An Introduction,* 2nd ed. (Frederick, Md.: Aspen Publishing, 2020), 260.

129 **"illegal dancing":** Michael Cooper, "Cabaret Law, Decades Old, Faces Repeal," *New York Times,* November 20, 2003.

129 **Maple was one of the first:** Jack Maple with Chris Mitchell, *The Crime Fighter: How You Can Make Your Community Crime Free* (New York: Broadway Books, 2000).

130 **The NYPD had:** Author's interview with NYPD financial crimes officer who participated on condition that he not be named, 2021. See also "NYPD Questions and Answers: Facial Recognition," https://www.nyc.gov/site/nypd/about/about-nypd/equipment-tech/facial-recognition.page.

130 **About an hour into the meeting:** Events as reflected in emails included in NYPD public records request, MuckRock.

130 **The app started generating:** Author's interview with Hoan Ton-That, 2021.

131 **Flanagan ran it up:** NYPD public records request, MuckRock.

131 **LOVEINT:** Siobhan Gorman, "NSA Officers Spy on Love Interests," *Wall Street Journal,* August 23, 2013.

131 **to local police officers:** Sadie Gurman, "Across US, Police Officers Abuse Confidential Databases," Associated Press, September 28, 2016.
131 **It was at that point:** NYPD public records request, MuckRock.
132 **a more regular office space:** Author's interview with Hoan Ton-That, 2021.
132 **helped the NYPD catch a pedophile:** Author's interview with Hoan Ton-That for *The New York Times,* 2020. Also Clearview promo packet, company document provided by Ton-That, on file with author.
132 **two on-duty cops:** Craig McCarthy, Tina Moore, and Bruce Golding, "Internal Affairs Used Facial Recognition Software Against NYPD Cops: Email," *New York Post,* April 11, 2021.
132 **Schwartz, meanwhile, was trying:** NYPD public records request, MuckRock.
133 **Ton-That wanted:** Author's interview with Hoan Ton-That, 2021.
133 **"It was definitely":** Author's interview with NYPD financial crimes officer, 2021.
133 **eleven thousand searches:** Ryan Mac et al., "Clearview AI Offered Thousands of Cops Free Trials," BuzzFeed News, April 6, 2021.
133 **"perception of it":** Author's interview with NYPD financial crimes officer, 2021.
133 **"But with child exploitation":** Ibid.
133 **an NYPD lieutenant introduced:** Indiana State Police public records request for emails about Clearview AI, filed by Anna Venarchik, February 2022, on file with author.
133 **$49,500 subscription:** Indiana State Police public records request for contracts, purchase orders, procurement records, agreements, invoices, or payment records related to Clearview AI, filed by author, December 2022. On file with author.
133 **"I am often asked":** Indiana State Police public records request for emails, 2022.
134 **One of the CrimeDex ads:** Public records request to the Seattle Police Department, filed by Bridget Brululo. "Clearview AI and SPD," MuckRock, November 3, 2020, https://cdn.muckrock.com/foia_files/2020/11/03/REL_P59603_Responsive_Emails_FINAL.pdf.
134 **But it was still:** Luke O'Brien, "The Far-Right Helped Create the World's Most Powerful Facial Recognition Technology," HuffPost, April 7, 2020.
134 **A college dropout:** "I dropped out of my very fancy college in America." "An Interview with Marko Jukic: Nobody Knows How to Learn a Language," HackerNoon, February 14, 2019.
134 **had worked with Charles Johnson:** Author's interview with Charles Johnson, 2021.
134 **"exercises in theatrical hyperbole":** O'Brien, "The Far-Right Helped Create the World's Most Powerful Facial Recognition Technology."
134 **spreading like wildfire:** Author's interview with David Scalzo, 2019.
134 **When a Seattle police detective:** This refers to an email from the Seattle Police Department in response to a public records request filed by Bridget Brululo. "Clearview AI and SPD," MuckRock, November 3, 2020.

134 **In May 2019:** I first reported about this case in *The New York Times Magazine:* Kashmir Hill, "Your Face Is Not Your Own," *New York Times Magazine,* March 18, 2021. Findley was unnamed in that piece, but he later revealed his identity in an onstage interview with Hoan Ton-That at a security conference. Hoan Ton-That, "Hoan Ton-That Interviews Josh Findley, Special Agent at U.S. Department of Homeland Security," Clearview AI, October 8, 2021.

135 **Google went a step further:** Nikola Todorovic and Abhi Chaudhuri, "Using AI to Help Organizations Detect and Report Child Sexual Abuse Material Online," Google, September 3, 2018, reported in Kashmir Hill, "A Dad Took Photos of His Naked Toddler for the Doctor. Google Flagged Him as a Criminal," *New York Times,* August 21, 2022.

136 **"the biggest breakthrough":** Kashmir Hill and Gabriel J. X. Dance, "Clearview's Facial Recognition App Is Identifying Child Victims of Abuse," *New York Times,* February 7, 2020.

136 **including Interpol:** Ryan Mac, Caroline Haskins, and Logan McDonald, "Clearview AI's Facial Recognition Tech Is Being Used by the Justice Department, ICE, and the FBI," BuzzFeed News, February 27, 2020; Ryan Mac et al., "See If Your Police Used Facial Recognition Tech Clearview AI," BuzzFeed News, April 10, 2021; Ryan Mac, Caroline Haskins, and Antonio Pequeño IV, "Clearview AI Offered Free Trials to Police Around the World," BuzzFeed News, August 25, 2021.

136 **Among the heaviest users:** Mac et al., "See If Your Police Used Facial Recognition Tech Clearview AI."

136 **Swedish Police Authority:** The Swedish Authority for Privacy Protection later fined its police approximately €250,000 for using Clearview in violation of federal law and ordered it to inform people whose photos were run.

137 **a marketing document:** Clearview AI Investor Presentation, December 2021, on file with author.

137 **Amazon press release:** "Amazon Rekognition Announces Real-Time Face Recognition, Text in Image Recognition, and Improved Face Detection," Amazon Web Services, Inc., https://aws.amazon.com/about-aws/whats-new/2017/11/amazon-rekognition-announces-real-time-face-recognition-text-in-image-recognition-and-improved-face-detection/.

137 **What worried Cagle:** Author's interview with Matt Cagle, 2022. Matt Cagle and Nicole A. Ozer, "Amazon Teams Up with Law Enforcement to Deploy Dangerous New Face Recognition Technology," American Civil Liberties Union of Northern California, May 22, 2018.

137 **discovered that the city of Orlando:** Cagle and Ozer, "Amazon Teams Up with Law Enforcement."

138 **ultimately abandoned:** Joey Roulette, "Orlando Cancels Amazon Rekognition Program, Capping 15 Months of Glitches and Controversy," *Orlando Weekly,* July 18, 2019.

138 **a tech-savvy colleague:** Author's interview with Jacob Snow, 2022.

138 **In the ACLU's test:** Jacob Snow, "Amazon's Face Recognition Falsely

Matched 28 Members of Congress with Mugshots," American Civil Liberties Union, July 26, 2018.

138 **launched a campaign:** Author's interviews with Kade Crockford, Matt Cagle, and Nicole Ozer, 2021.

139 **"It rubbed people":** Author's interview with a person at the DAGA event who spoke on the condition of not being named, 2020.

139 **Tong said that the demo:** Tong's comments about the episode to author at an NAAG conference in Vermont, 2021.

Chapter 14: "You Know What's Really Creepy?" (2011–2019)

140 **he turned to:** Author's interview with Alvaro Bedoya, 2021.

140 **"I literally got stuck":** Ibid.

140 **"What They Know" series:** Julia Angwin, et al., "What They Know" series, *Wall Street Journal,* 2010–2012, https://www.wsj.com/news/types/what-they-know.

141 **so full that they had to:** "I am sorry that everyone was not able to get into the room, into the hearing room, but we are streaming live on C-SPAN, thankfully," said Al Franken at the hearing. *Protecting Mobile Privacy: Your Smartphones, Tablets, Cell Phones and Your Privacy, Hearing Before the Subcommittee on Privacy, Technology and the Law of the Committee on the Judiciary,* United States Senate, 112th Cong., May 10, 2011, https://www.judiciary.senate.gov/imo/media/doc/CHRG-112shrg86775.pdf, i.

141 **Al Franken sat front:** "Mobile Technology and Privacy, Government Officials," C-SPAN, May 10, 2011.

141 **"When I was growing up":** *Protecting Mobile Privacy,* 3.

141 **"Maybe smart phones are":** Matt Jaffe, "Senate Panel Grills Apple, Google over Cell Phone Privacy," ABC News, May 10, 2011.

141 **"Hey, Alvaro":** Conversation as recalled by Alvaro Bedoya in interview with the author, 2021.

142 **"He would tell a funny story":** Ibid.

142 **Microsoft offering facial recognition:** "Kinect Fact Sheet," Xbox & Kinect press materials, Microsoft, June 14, 2010. On file with author.

142 **turning it on for everybody:** "Making Photo Tagging Easier," The Facebook Blog, December 15, 2010, https://web.archive.org/web/20110107154043/, http://blog.facebook.com/blog.php?post=467145887130; Ben Parr, "Facebook Brings Facial Recognition to Photo Tagging," CNN, December 16, 2010.

142 **"super-creepy":** Sarah Jacobsson Purewal, "Why Facebook's Facial Recognition Is Creepy," *PCWorld,* June 8, 2011.

142 **as opposed to in Europe:** Somini Sengupta and Kevin J. O'Brien, "Facebook Can ID Faces, but Using Them Grows Tricky," *New York Times,* September 21, 2012; Helen Pidd, "Facebook Facial Recognition Software Violates Privacy Laws, Says Germany," *The Guardian,* August 3, 2011.

142 **forced all of its users:** Ryan Singel, "Public Posting Now the Default on Facebook," *Wired,* December 9, 2009.

143 **including the ACLU:** "Before the recent changes, you had the option of exposing only a 'limited' profile, consisting of as little as your name and networks, to other Facebook users—and nothing at all to Internet users at large. . . . Now your profile picture, current city, friends list, gender, and fan pages are 'publicly available information.'" Nicole Ozer, "Facebook Privacy in Transition—But Where Is It Heading?," American Civil Liberties Union, December 9, 2009.

143 **Facebook initially refused:** Author's interview with Alvaro Bedoya, 2021.

143 **Facebook had just acquired:** Alexia Tsotsis, "Facebook Scoops Up Face.com for $55–60M to Bolster Its Facial Recognition Tech," TechCrunch, June 18, 2012.

143 **Bedoya assured the company:** Author's interview with Alvaro Bedoya, 2021.

143 **"People think hearings":** Ibid.

143 **"I want to be clear":** *What Facial Recognition Technology Means for Privacy and Civil Liberties, Hearing Before the Subcommittee on Privacy, Technology, and the Law of the Committee on the Judiciary, United States Senate,* 112th Cong., July 18, 2012, https://www.govinfo.gov/content/pkg/CHRG-112shrg86599/pdf/CHRG-112shrg86599.pdf, 1.

143 **"Ordinary, honest people":** Larry Amerson, ibid., 34.

144 **"random strangers":** Rob Sherman, ibid., 23.

144 **Behind the scenes:** Author's interview with Alvaro Bedoya, 2021.

144 **"I was unequivocally told":** Ibid.

144 **"Once you generate a faceprint":** *What Facial Recognition Technology Means for Privacy and Civil Liberties,* 35.

144 **"five or ten years down the road":** Rob Sherman, ibid., 35–36.

144 **Tommer Leyvand sat:** Video of the demo. On file with author.

145 **"It'll say the name":** Ibid.

145 **"That's me":** Ibid.

145 **"supposed to be a joke":** Author's interview with former Facebook employee who spoke on condition that he not be named, 2021.

145 **"direct brain interface":** Josh Constine, "Facebook Is Building Brain-Computer Interfaces for Typing and Skin-Hearing," TechCrunch, April 19, 2017.

145 **smart glasses:** Mike Isaac, "Smart Glasses Made Google Look Dumb. Now Facebook Is Giving Them a Try," *New York Times,* September 9, 2021.

146 **shortly after his arrival:** Author's interview with Yaniv Taigman, 2021.

146 **"We told him":** Ibid.

146 **"Let's go to one million":** Mark Zuckerberg's response as recalled by Yaniv Taigman in ibid.

146 **"No way":** Yaniv Taigman's recollection of the conversation in ibid.

146 **blew the competition:** The SuperVision developers were Geoffrey Hinton and his two graduate students, Ilya Sutskever and Alex Krizhevsky. For

more on this, see Cade Metz, *Genius Makers: The Mavericks Who Brought AI to Google, Facebook, and the World* (New York: Dutton, 2021).

147 **Taigman realized:** Author's interview with Yaniv Taigman, 2021.

147 **4.4 million photos:** Yaniv Taigman et al., "DeepFace: Closing the Gap to Human-Level Performance in Face Verification," *2014 IEEE Conference on Computer Vision and Pattern Recognition,* June 2014, 1701–08.

147 **had been chosen "randomly":** Author's interview with Yaniv Taigman, 2021.

147 **"I was sure":** Ibid.

148 **"creepy":** Dino Grandoni, "Facebook's New 'DeepFace' Program Is Just as Creepy as It Sounds," Huffington Post, March 14, 2014.

148 **"scary good":** Will Oremus, "Facebook's New Face-Recognition Software Is Scary Good," Slate, March 18, 2014.

148 **"This is theoretical research":** Ibid.

148 **dramatically improving its ability:** Taylor Hatmaker, "Facebook Will Tag Every Photo Ever Taken of You—Whether You Like It or Not," The Daily Dot, February 7, 2015.

148 **Taigman and his team were relocated:** Author's interview with Yaniv Taigman, 2021.

148 **"Everything new is scary":** Ibid.

148 **technical sweetness:** Author's interview with Heather Douglas, 2022.

149 **created a real sense of momentum:** Author's interview with Alvaro Bedoya, 2021.

149 **The bill passed out of:** S.1223: Location Privacy Protection Act of 2012, 112th Congress (2011–2012), 112th Cong., https://www.congress.gov/112/bills/s1223/BILLS-112s1223rs.pdf.

149 **Bedoya felt the wind:** Author's interview with Alvaro Bedoya, 2021.

149 **Glenn Greenwald had published:** Glenn Greenwald, "NSA Collecting Phone Records of Millions of Verizon Customers Daily," *The Guardian,* June 6, 2013.

149 **"Al, we've got to investigate":** Alvaro Bedoya's recollection of the conversation in interview with the author, 2021.

149 **"The landscape changed":** Author's interview with Alvaro Bedoya, 2021.

150 **"This bill is going":** Email to Alvaro Bedoya, March 27, 2014. On file with author.

150 **industry was coalescing to kill the bill:** Author's interview with Alvaro Bedoya, 2021.

150 **"helped prevent potentially damaging":** State Privacy and Security Coalition Form 990, filed with the Internal Revenue Service, 2012, https://projects.propublica.org/nonprofits/organizations/262149959.

150 **"They were nice to your face":** Author's interview with Alvaro Bedoya, 2021.

151 **The first seven:** "Despite years of inactivity under Illinois BIPA, seven cases were filed in 2015; plaintiffs then filed seven more putative class actions in 2016. The cases filed in 2015 and 2016 generally targeted retailers and online service providers, alleging that they improperly collected and stored photographs." Carley Daye Andrews et al., "Litigation Under Illinois Biometric

Information Privacy Act Highlights Biometric Data Risks," K&L Gates, November 7, 2017.

151 **One of them was against Facebook:** *Carlo Licata, Adam Pezen and Nimesh Patel v. Facebook,* Consolidated Class Action Complaint, United States District Court, Northern District of California, filed August 28, 2015.

152 **Six Flags gave up and settled:** Jonathan Bilyk, "Six Flags Inks $36M Deal to End Fingerprint Scan Class Action That Resulted in Landmark IL Supreme Court Decision," *Cook County Record,* June 8, 2021.

152 **Turk provided an expert opinion:** Turk's expert testimony was withheld from the public at Facebook's request for proprietary reasons, but parts of it were cited in the judge's order on summary judgment. *In re Facebook Biometric Information Privacy Litigation,* United States District Court, Northern District of California, filed May 29, 2018.

152 **Facebook agreed to pay:** Jennifer Bryant, "Facebook's $650M BIPA Settlement 'a Make-or-Break Moment,'" International Association of Privacy Professionals, March 5, 2021.

152 **"machine perception":** Tommer Leyvand's LinkedIn profile, https://www.linkedin.com/in/tommerl.

Chapter 15: Caught in a Dragnet

153 **stemmed from his own experience:** Author's interview with Freddy Martinez, 2020.

153 **money had never come easily:** Author's interview with Freddy Martinez, 2021.

153 **In 2012:** John McCrank, "Knight Capital Gets $400 Million Rescue, Shares Tumble," Reuters, August 6, 2012; Jacob Bunge, Anupreeta Das, and Telis Demos, "Knight Capital Gets Lifeline," *Wall Street Journal,* August 7, 2012; "SEC Charges Knight Capital with Violations of Market Access Rule," U.S. Securities and Exchange Commission, October 16, 2013, https://www.sec.gov/news/press-release/2013-222.

153 **"When you have money":** Author's interview with Freddy Martinez, 2021.

154 **large-scale demonstrations:** "Protesters Rally Outside NATO Summit After March from Grant Park," CBS Chicago, May 20, 2012.

154 **Martinez decided:** Author's interview with Freddy Martinez, 2020.

154 **a large group of police officers:** Freddy Martinez's recollection of the episode in interviews with the author, 2020–2021.

154 **Martinez thought:** Author's interview with Freddy Martinez, 2021.

154 **"It just looked so funny":** Author's interview with Freddy Martinez, 2021.

154 **several people wound up:** Adam Gabbatt, "Chicago Police Clash with Nato Summit Protesters," *The Guardian,* May 21, 2012.

154 **a device called a stingray:** Though it had been secretly used by police for years, the stingray device had only recently been exposed to the public. Jennifer Valentino-DeVries, "FBI's 'Stingray' Cellphone Tracker Stirs a Fight over Search Warrants, Fourth Amendment," *Wall Street Journal,* September

22, 2011. See also Cyrus Farivar, "How a Hacker Proved Cops Used a Secret Government Phone Tracker to Find Him," Politico Magazine, June 3, 2018.

155 **national news story:** John Dodge, "After Denials, Chicago Police Department Admits Purchase of Cell-Phone Spying Devices," CBS Chicago, October 1, 2014; Matt Richtel, "A Police Gadget Tracks Phones? Shhh! It's Secret," *New York Times,* March 16, 2015.

155 **described him as "fidgeting":** Justin Glawe, "Freddy Martinez Is Exposing Chicago Cops' NSA-Style Surveillance Gear," *VICE,* March 31, 2015.

155 **"the sweet spot":** *Facial Recognition Technology: Part I: Its Impact on Our Civil Rights and Liberties, Hearing before the Committee on Oversight and Reform, House of Representatives,* 116th Cong., May 22, 2019, https://docs.house.gov/meetings/GO/GO00/20190522/109521/HHRG-116-GO00-Transcript-20190522.pdf, 29.

156 **She had discovered:** Joy Buolamwini, "How I'm Fighting Bias in Algorithms" (video), TEDxBeaconStreet, March 9, 2017.

156 **investigated products:** Joy Buolamwini, "Gender Shades: Intersectional Phenotypic and Demographic Evaluation of Face Datasets and Gender Classifiers," MIT Media Lab, September 2017.

156 **"Our faces may well be":** *Facial Recognition Technology: Part I: Its Impact on Our Civil Rights and Liberties,* 5.

156 **"private companies from collecting":** Written testimony of Joy Buolamwini, May 22, 2019, https://www.congress.gov/116/meeting/house/109521/witnesses/HHRG-116-GO00-Wstate-BuolamwiniJ-20190522.pdf.

156 **Garvie had been one:** Author's interview with Alvaro Bedoya, 2021; author's interview with Clare Garvie, 2021.

156 **In a 2016 report:** Clare Garvie, Alvaro M. Bedoya, and Jonathan Frankle, "The Perpetual Line-up: Unregulated Police Face Recognition in America," Georgetown Law Center on Privacy & Technology, October 18, 2016.

156 **Garvie had dropped:** Clare Garvie, "Garbage In. Garbage Out. Face Recognition on Flawed Data," report published by Georgetown Law Center on Privacy & Technology, May 16, 2019.

156 **"My research has uncovered":** Statement by Clare Garvie, *Facial Recognition Technology: Part I: Its Impact on Our Civil Rights and Liberties,* 8.

157 **"Face recognition is already":** Written testimony of Clare Garvie, May 22, 2019, https://www.congress.gov/116/meeting/house/109521/witnesses/HHRG-116-GO00-Wstate-GarvieC-20190522.pdf.

157 **"queen of the FOIA freaks":** Beryl Lipton, bio, Twitter, https://twitter.com/_blip_.

157 **the first words Martinez saw:** Author's interview with Freddy Martinez, 2021.

157 **"In the simplest terms":** The documents that Freddy Martinez obtained in the public records request are available at "Facial Recognition—Atlanta (GA)," MuckRock, October 3, 2019.

158 **the FBI had only:** Statement by Kimberly Del Greco of the FBI's Criminal Justice Information Services Division at a congressional hearing. *Facial*

Recognition Technology: Part II: Ensuring Transparency in Government Use, Hearing before the Committee on Oversight and Reform, House of Representatives, 116th Cong., June 4, 2019, https://www.govinfo.gov/content/pkg/CHRG-116hhrg36829/pdf/CHRG-116hhrg36829.pdf.

Chapter 16: Read All About It

160 **I later learned:** Author's interviews with Lisa Linden and Hoan Ton-That, 2021.

160 **"I've had to significantly":** Richard Schwartz, email to author, December 8, 2019.

160 **Ton-That saw via Facebook:** Author's interview with Keith Dumanski, 2021.

160 **Ton-That hoped:** Author's interview with Hoan Ton-That, 2021.

161 **that Ton-That had drafted:** Author's interview with Keith Dumanski, 2021.

161 **During Schwartz's days:** Author's interview with Lisa Linden, 2021.

161 **Schwartz and Ton-That trekked:** Author's interviews with Lisa Linden and Hoan Ton-That, 2021.

161 **Linden broke the news:** Author's interviews with Linden and Hoan Ton-That, 2021.

161 **Ton-That spent hours:** Ibid.

162 **"When you're doing something":** Author's interview with Hoan Ton-That for *The New York Times,* 2020.

162 **"Every technology":** Ibid.

162 **"It must have been a bug":** Ibid.

163 **"We had a system":** Hoan Ton-That, speaking with journalist Rolfe Winkler, *WSJ Tech Live,* October 20, 2020.

163 **A congressional staffer:** Author's interview with a source who spoke on the condition that he not be named, 2021.

163 **"sketchy":** Author's interview with Charles Johnson, 2021.

163 **Now Ton-That demonstrated:** Author's interview with Hoan Ton-That for *The New York Times,* 2020.

164 **"I have to think about that":** Ibid.

164 **"They're saying it's crazy":** Conversation as recalled by Hoan Ton-That and Lisa Linden in interviews with author, 2021.

164 **Ton-That decided to subscribe:** Author's interview with Hoan Ton-That, 2021.

164 **It was on the front page:** Kashmir Hill, "The Secretive Company That Might End Privacy as We Know It," *New York Times,* January 18, 2020.

164 **"in our labs":** Author's email exchanges with Hoan Ton-That and Lisa Linden, 2020.

164 **People recognized him at parties:** Author's interview with Hoan Ton-That, 2021.

164 **"I'm walking clickbait":** Author's interview with Hoan Ton-That, 2020.

165 **"Widespread use of your technology":** Edward Markey to Hoan Ton-That; Ed Markey, "Senator Markey Queries Clearview About Its Facial Recognition Product and Sale to Law Enforcement," January 23, 2020, https://www.markey.senate.gov/news/press-releases/senator-markey-queries-clearview-about-its-facial-recognition-product-and-sale-to-law-enforcement.

165 **he was "troubled":** Author's interview with Gurbir Grewal; Kashmir Hill, "New Jersey Bars Police from Using Clearview Facial Recognition App," *New York Times,* January 25, 2020.

165 **Apple revoked:** Logan McDonald, Ryan Mac, and Caroline Haskins, "Apple Just Disabled Clearview AI's iPhone App for Breaking Its Rules on Distribution," BuzzFeed News, February 28, 2020.

165 **a federal court had declared:** Camille Fischer and Andrew Crocker, "Victory! Ruling in hiQ v. Linkedin Protects Scraping of Public Data," Electronic Frontier Foundation, September 10, 2019.

165 **dug into the company's ties:** Luke O'Brien, "The Far-Right Helped Create the World's Most Powerful Facial Recognition Technology," HuffPost, April 7, 2020.

PART III: FUTURE SHOCK

Chapter 17: "Why the Fuck Am I Here?" (2020)

169 **Melissa Williams got a call:** Author's interview with Robert and Melissa Williams for *The New York Times,* 2020. Kashmir Hill, "Wrongfully Accused by an Algorithm," *New York Times,* June 24, 2020.

169 **"He needs to turn himself in":** The conversation with the officer as recalled by Melissa Williams, ibid.

169 **"I just got":** Their conversation as recalled by Robert and Melissa Williams, ibid.

170 **"You need to come":** The conversation with the officer as recalled by Robert Williams, ibid.

170 **"Send Robert out":** The interaction as recalled by Melissa Williams, ibid.

170 **"You have visitors here":** Their conversation as recalled by Robert and Melissa Williams, ibid.

171 **The squad car pulled up:** Robert Williams's arrest was filmed by the two officers on their body cameras. The description of events and dialogue is based on that footage. On file with author.

172 **Robert seethed:** Author's interview with Robert and Melissa Williams for *The New York Times,* 2020.

173 **None of the neighbors:** Ibid.

174 **"Where did Daddy go?":** As recalled by Melissa Williams, ibid.

174 **"You have a nice car":** Interaction in car as recalled by Robert Williams, ibid.

174 ***Yeah, yeah, yeah:*** Ibid.

174 **"What are you in for?":** Robert Williams's recollection of his time in detention, ibid.

175 **There was a table and three chairs:** Recording of custodial interrogation of Robert Williams by Detectives James Ronan and Benjamin Atkinson, January 10, 2020.

175 **"I don't even know":** The description of actions and dialogue during the interrogations is based on the video recording, ibid.

175 **This was not true:** Elie Mystal, *Allow Me to Retort: A Black Man's Guide to the Constitution* (New York: The New Press, 2022).

177 **According to a police report:** The description of the crime is based on case documents on file with author.

177 **two different facial recognition algorithms:** Author's interview with a manager at DataWorks Plus for *The New York Times.* Hill, "Wrongfully Accused by an Algorithm."

177 **The year prior:** Patrick Grother, "Face Recognition Vendor Test Part 3: Demographic Effects," National Institute of Standards and Technology, December 2019.

178 **The media takeaway:** Natasha Singer and Cade Metz, "Many Facial-Recognition Systems Are Biased, Says U.S. Study," *New York Times,* December 19, 2019.

178 **there were good algorithms:** Author's interview with Clare Garvie, 2020.

178 **web cameras:** wzamen01, "HP Computers Are Racist," YouTube, December 11, 2009, https://www.youtube.com/watch?v=t4DT3tQqgRM.

178 **image recognition:** Jacob Brogan, "Google Scrambles After Software IDs Photo of Two Black People as 'Gorillas,'" Slate, June 30, 2015.

178 **soap dispensers:** Chukwuemeka Afigbo, "If you have ever had a problem grasping the importance of diversity in tech and its impact on society, watch this video," Twitter, August 16, 2017, https://twitter.com/nke_ise/status/897756900753891328.

179 **Kodak had distributed:** Mandalit del Barco, "How Kodak's Shirley Cards Set Photography's Skin-Tone Standard," NPR, November 13, 2014.

179 **Problems persist:** Xavier Harding, "Keeping 'Insecure' Lit: HBO Cinematographer Ava Berkofsky on Properly Lighting Black Faces," Mic, September 6, 2017.

179 **Chinese companies that made deals:** Amy Hawkins, "Beijing's Big Brother Tech Needs African Faces," *Foreign Policy,* July 24, 2018; Zhang Hongpei, "Chinese Facial ID Tech to Land in Africa," *Global Times,* May 17, 2018, https://web.archive.org/web/20180625130452/http:/www.globaltimes.cn/content/1102797.shtml.

179 **In 2019, Google asked:** Sean Hollister, "Google Contractors Reportedly Targeted Homeless People for Pixel 4 Facial Recognition," The Verge, October 3, 2019.

179 **ensure its Face Unlock feature:** Jack Nicas, "Atlanta Asks Google Whether It Targeted Black Homeless People," *New York Times*, October 4, 2019.

180 **A digital image examiner:** Case documents. On file with author.

180 **A former parole agent:** Jennifer Coulson, speaker bio, Facial Recognition and

Privacy Workshop, CSU College of Law, March 29, 2019, https://www.law.csuohio.edu/newsevents/events/2019032907305197.

180 **243 photos:** Anderson Cooper, "Police Departments Adopting Facial Recognition Tech amid Allegations of Wrongful Arrests," *60 Minutes,* May 16, 2021, https://www.cbsnews.com/news/facial-recognition-60-minutes-2021-05-16/.

181 **they offered Robert:** Robert Williams's recollection of his time in jail, 2020.

182 **"I'm innocent":** Conversations with lawyers as recalled by Robert Williams in interview with the author, 2020.

182 **"This does not in any way":** County of Wayne Office of the Prosecuting Attorney, "WCPO Statement in Response to New York Times Article Wrongfully Accused by an Algorithm," June 24, 2020, https://int.nyt.com/data/documenthelper/7046-facial-recognition-arrest/5a6d6d0047295fad363b/optimized/full.pdf#page=1.

183 **"sloppy":** Author's interview with Detroit police chief James Craig for *The New York Times,* 2020.

183 **But no one called:** Author's interview with Robert and Melissa Williams, 2021.

183 **The stress of what had happened:** Ibid.

183 **a second case:** Elisha Anderson, "Controversial Detroit Facial Recognition Got Him Arrested for a Crime He Didn't Commit," *Detroit Free Press,* July 10, 2020.

183 **a third case:** Kashmir Hill, "Flawed Facial Recognition Leads to Arrest and Jail for New Jersey Man," *New York Times,* December 29, 2020.

183 **a fourth man:** Khari Johnson, "Face Recognition Software Led to His Arrest. It Was Dead Wrong," *Wired,* February 28, 2023.

183 **then a fifth:** John Simerman, "JPSO Used Facial Recognition Technology to Arrest a Man. The Tech Was Wrong," NOLA.com, January 2, 2023.

183 **was Clearview AI:** Kashmir Hill and Ryan Mac, "Police Relied on Hidden Technology and Put the Wrong Person in Jail," *New York Times,* March 31, 2023.

183 **More than a year:** Author's interview with Robert and Melissa Williams, 2021.

184 **"I don't know when":** Ibid.

Chapter 18: A Different Reason to Wear a Mask

186 **In March 2020:** Kirsten Grind, Robert McMillan, and Anna Wilde Mathews, "To Track Virus, Governments Weigh Surveillance Tools That Push Privacy Limits," *Wall Street Journal,* March 17, 2020.

186 **reports emerged that South Korea:** Sangmi Cha, "S.Korea to Test AI-Powered Facial Recognition to Track COVID-19 Cases," Reuters, December 13, 2021.

186 **China:** "Facial Recognition Tech Fights Coronavirus in Chinese City," France 24, July 13, 2021.

186 **Russia:** "Moscow Deploys Facial Recognition Technology for Coronavirus Quarantine," Reuters, February 21, 2020.

186 **After traveling abroad:** Patrick Reevell, "How Russia Is Using Facial Recognition to Police Its Coronavirus Lockdown," ABC News, April 30, 2020.

186 **tried to pass a bill:** Ed Chau, "Facial Recognition Regulation: AB 2261 Is a Long Overdue Solution," CalMatters, June 2, 2020.

186 **But that bill:** Ryan Johnston, "Facial Recognition Bill Falls Flat in California Legislature," StateScoop, June 4, 2020.

187 **Richard Schwartz privately circulated:** Email and slideshow. On file with author.

187 **One of the companies:** Ryan Mac, Caroline Haskins, and Logan McDonald, "Clearview's Facial Recognition App Has Been Used by the Justice Department, ICE, Macy's, Walmart, and the NBA," BuzzFeed News, February 28, 2020.

187 **the Garden put thousands of lawyers:** Kashmir Hill and Corey Kilgannon, "Madison Square Garden Uses Facial Recognition to Ban Its Owner's Enemies," *New York Times,* December 22, 2022.

188 **David Scalzo, meanwhile, was frustrated:** Author's interview with David Scalzo, 2021.

188 **"They're limiting the eyeballs":** Ibid.

188 **Ton-That and Schwartz told him:** David Scalzo's recollection in interview with the author, 2021.

188 **"The delta between":** Author's interview with David Scalzo, 2021.

Chapter 19: I Have a Complaint

190 **was deeply troubled:** Author's interview with Matthias Marx, 2021.

190 **"Dear sir or madam":** Matthias Marx, email to privacy-requests@clearview.ai, January 20, 2020. On file with author.

190 **European law required:** Article 15 of the European General Data Protection Regulation (GDPR): Right of Access by the Data Subject, https://gdpr-text.com/read/article-15/.

191 **But Marx did not think:** Author's interview with Matthias Marx, 2021.

191 **"Data Access Request Complete":** Face Search Results, report prepared February 18, 2020. On file with author.

192 **Marx wrote a complaint:** Matthias Marx and Alan Dahi, "Clearview AI and the GDPR," November 7, 2020, https://marx.wtf/2020-11-FnF-Clearview.pdf; Patrick Beuth, "Hamburgs Datenschützer leitet Prüfverfahren gegen Clearview ein," *Der Spiegel,* March 25, 2020.

192 **local privacy agency:** "Auskunftsheranziehungsbescheid gegen Clearview AI erlassen—Transparente Antworten zum Datenschutz gefordert!," Hamburg Commissioner for Data Protection and Freedom of Information, August 18, 2020.

192 **Clearview sent Marx another:** privacy@clearview.ai, email to Matthias Marx, May 18, 2020. On file with author.

192 **which ruled:** Author's email exchange with Martin Schemm, press officer for the Hamburg Commissioner for Data Protection and Freedom of Information, December 15, 2021.

193 **the cohort of international privacy regulators:** "Clearview AI's Unlawful Practices Represented Mass Surveillance of Canadians, Commissioners Say," Office of the Privacy Commissioner of Canada, February 3, 2021; "Clearview AI Breached Australians' Privacy," Office of the Australian Information Commissioner, November 3, 2021; "ICO Issues Provisional View to Fine Clearview AI Inc over £17 Million," Information Commissioner's Office (UK), November 29, 2021; "Facial Recognition: Italian SA Fines Clearview AI EUR 20 Million," European Data Protection Board, March 10, 2022; "Hellenic DPA Fines Clearview AI 20 Million Euros," European Data Protection Board, July 20, 2022; "Facial Recognition: 20 Million Euros Penalty against CLEARVIEW AI," CNIL, October 20, 2022.

193 **"What Clearview does":** "Clearview AI's Unlawful Practices Represented Mass Surveillance of Canadians, Commissioners Say."

193 **"I grew up in Australia":** Hoan Ton-That, "Personal Statement on the Australian Privacy Commission"; Lisa Linden, email to author, November 2, 2021.

193 **caught a mobster:** Elisabetta Povoledo, "Italian Mobster, 16 Years on the Lam, Found Working at a Pizzeria," *New York Times,* February 3, 2023.

194 **Ton-That shrugged off:** Author's interviews with Hoan Ton-That, 2021–2022.

194 **It had a patent:** Richard Bilbao, "Disney Patent Wants to Track Guests by Scanning Their Feet," *Orlando Business Journal,* July 27, 2016.

194 **He was utterly unsurprised:** "A couple of weeks ago, the #clearviewai reporting by @kashhill shook the world. Now we show how easy it is to download bulk-scale data from Instagram—and how advanced open source #facerec technology already is," Twitter, https://twitter.com/grssnbchr/status/1225846790093725696. See also author's interview with Timo Grossenbacher, 2020.

194 **Anyone could do this:** Author's interview with Timo Grossenbacher, 2020. See also Timo Grossenbacher and Felix Michel, "Automatische Gesichtserkennung—So einfach ist es, eine Überwachungsmaschine zu bauen" ("It's That Easy to Build a Surveillance Machine"), SRF Schweizer Radio und Fernsehen, February 7, 2020.

195 **barely batted an eye:** "They would block us for an hour but then they would let us start downloading again." Ibid.

195 **"Oh my God":** Ibid.

195 **The Swiss data commissioner:** Author's interview with Timo Grossenbacher, 2020.

195 **"aren't rocket scientists":** Ibid.

Chapter 20: The Darkest Impulses

197 **David had an addiction:** Author's interview with source who spoke on the condition that he not be named, 2021.

198 **"You find them":** Ibid.

198 **In 2014, a freshman:** Alex Morris, "The Blue Devil in Miss Belle Knox: Meet Duke Porn Star Miriam Weeks," *Rolling Stone,* April 23, 2014.

198 **"I was called a 'slut'":** Miriam Weeks, "I'm the Duke University Freshman Porn Star and for the First Time I'm Telling the Story in My Words," xoJane, February 21, 2014, Wayback Machine, https://web.archive.org/web/20140301125826/http://www.xojane.com/sex/duke-university-freshman-porn-star.

199 **"We're not a protected class":** Author's interview with Ela Darling, 2021.

199 **He considered himself:** Author's interview with David, 2021.

199 **"I went through my entire":** Ibid.

200 **"It turned out she shot porn":** Ibid.

200 **"PROtect plan":** "Pricing," PimEyes, PimEyes.com/en/premium. "Managing current and future results" is one of the PROtect features.

200 **professionalized sextortion:** Author's interview with Cher Scarlett for *The New York Times.* Kashmir Hill, "A Face Search Engine Anyone Can Use Is Alarmingly Accurate," *New York Times,* May 26, 2022.

200 **"hacker" types:** Author's interview with Giorgi Gobronidze for *The New York Times*, 2022.

201 **"Ignorance is bliss":** Author's interview with David, 2021.

Chapter 21: Code Red (or, Floyd Abrams v. the ACLU)

202 **they immediately began thinking:** Author's interview with Nathan Wessler and Vera Eidelman, 2022.

202 **"The Dawn of Robot Surveillance":** Jay Stanley, "The Dawn of Robot Surveillance: AI, Video Analytics, and Privacy," American Civil Liberties Union, June 2019.

203 **"Something that people":** Author's interview with Nathan Wessler and Vera Eidelman, 2022.

203 **"Here was the Silicon Valley":** Ibid.

204 **"The question was":** Ibid.

204 **"Almost a century ago":** *David Mutnick v. Clearview AI, Inc., Richard Schwartz, and Hoan Ton-That,* Class Action Complaint, United States District Court, Northern District of Illinois, Eastern Division, filed January 22, 2020.

205 **"nightmare scenario":** Davey Alba, "A.C.L.U. Accuses Clearview AI of Privacy 'Nightmare Scenario,'" *New York Times,* May 28, 2020.

205 **Its main objective:** Author's interview with Nathan Wessler and Vera Eidelman, 2022.

205 **In July 2020:** Author's interview with Floyd Abrams, 2020.

206 **Schwartz explained:** Ibid.

206 **computer code:** See *Bernstein v. Department of Justice,* a U.S. Court of Appeals ruling in the 1990s in favor of Daniel Bernstein, a mathematics graduate student who wanted to publish a paper about an encryption algorithm that anyone could use to protect their data from the prying eyes of hackers or government agents. The government tried to restrict its publication, saying that the student was exporting a "munition" and would need to register as an "arms dealer." Bernstein was successfully represented by the Electronic Frontier Foundation.

206 **political spending:** The Supreme Court determined in the hotly contested *Citizens United v. Federal Election Commission* case in 2010 that the government couldn't limit how much money corporations and nonprofits spent on campaign advertising. Floyd Abrams represented Senator Mitch McConnell on the winning side of that argument.

206 **video games:** In 2011, in *Brown v. Entertainment Merchants Association,* the Supreme Court found a California law prohibiting the sale of violent video games to minors unconstitutional. The High Court said that violence didn't fit into an "obscenity" exception to the First Amendment because it applies only to sexual conduct and noted that many tales for children are violent. "Grimm's Fairy Tales, for example, are grim indeed," Justice Antonin Scalia wrote.

206 **web search results:** Several companies have sued Google to get the search engine to change their results, but judges have found again and again that the California company is entitled to rank websites as it chooses. In a case from 2003, an advertising company named Search King sued Google after a drop in its appearance in search rankings. A federal judge in Oklahoma dismissed the case, citing Google's First Amendment rights; Google's search results amounted to "constitutionally protected opinions."

206 **"I found it really interesting":** Author's interview with Floyd Abrams, 2020.

206 **"It took me a few days":** Ibid.

207 **After being tear-gassed:** Kashmir Hill, "Activists Turn Facial Recognition Tools Against the Police," *New York Times,* October 21, 2020.

208 **the solid gold AAA ratings:** David Segal, "A Matter of Opinion?," *New York Times,* July 18, 2009.

208 **In 2009:** Duff Wilson, "Tobacco Companies Sue to Loosen New Limits," *New York Times,* August 31, 2009.

209 **article I was working on:** Kashmir Hill, "Your Face Is Not Your Own," *New York Times,* March 18, 2021.

211 **"indisputable proposition":** Transcript of proceedings, *American Civil Liberties Union v. Clearview AI,* Circuit Court of Cook County, Illinois, April 2, 2021. On file with author.

211 **"What we're":** Ibid.

211 **"We're not arguing":** Ibid.

212 **"The fact that something":** Memorandum opinion and order, *American Civil*

Liberties Union v. Clearview AI, Circuit Court of Cook County, Illinois, August 27, 2021.

213 **In May 2022:** Ryan Mac and Kashmir Hill, "Clearview AI Settles Suit and Agrees to Limits on Facial Recognition Database," *New York Times,* May 9, 2022.

213 **organized an appreciation event:** The author attended the event on June 9, 2022.

Chapter 22: The Future Is Unevenly Distributed

214 **On that Thursday:** Marie Jackson and Mary O'Connor, "Covid: Face Mask Rules and Covid Passes to End in England," BBC News, January 18, 2022.

214 **The police parked:** The description of this facial recognition deployment is based on photos and videos taken by Silkie Carlo and Big Brother Watch, on file with author. They are also available on Twitter: Big Brother Watch, "Live Facial Recognition Cameras are being used by @metpoliceuk outside #OxfordCircus tube station today," January 28, 2022, https://twitter.com/BigBrotherWatch/status/1487073214345973768.

214 **had compiled a watch list:** "MPS LFR Deployments 2020–Date," Metropolitan Police, https://www.met.police.uk/SysSiteAssets/media/downloads/force-content/met/advice/lfr/deployment-records/lfr-deployment-grid.pdf.

215 **"helps the Metropolitan Police Service":** "MPS LFR Policy Document: Direction for the MPS Deployment of Overt Live Facial Recognition Technology to Locate Person(s) on a Watchlist," Metropolitan Police, November 29, 2022, https://www.met.police.uk/SysSiteAssets/media/downloads/force-content/met/advice/lfr/policy-documents/lfr-policy-document.pdf.

215 **It was the fourteenth time:** Ten deployments are recorded in this report: "Metropolitan Police Service Live Facial Recognition Trials," National Physical Laboratory / Metropolitan Police Service, trials period: August 2016–February 2019," February 2020, https://www.met.police.uk/SysSiteAssets/media/downloads/central/services/accessing-information/facial-recognition/met-evaluation-report.pdf. The remainder are described in "MPS LFR Deployments 2020–Date."

215 **the police tweeted:** MPS Westminster, "We'll be using Live Facial Recognition technology at key locations in Westminster on Friday 28 January from 12 noon. Find out more at http://met.police.uk/advice/advice-and-information/fr/facial-recognition," Twitter, January 28, 2022, https://twitter.com/mpswestminster/status/1487017790867394563?lang=en.

215 **those tweets were like:** Author's interview with Silkie Carlo, 2022.

215 **polls regularly found:** Vian Bakir et al., "Public Feeling on Privacy, Security and Surveillance: A Report by DATA-PSST and DCSS," November 2015, https://orca.cardiff.ac.uk/id/eprint/87335/1/Public-Feeling-on-Privacy-Security-Surveillance-DATAPSST-DCSS-Nov2015.pdf; Joel Rogers de Waal, "Security Trumps Privacy in British Attitudes to Cyber-Surveillance," YouGov, June 12, 2017; "Beyond Face Value: Public Attitudes to Facial Recognition

Technology," Ada Lovelace Institute, September 2, 2019 (70 percent of those polled approved of the use of facial recognition by police).

217 **Carlo's big fear:** Author's interview with Silkie Carlo, 2022.

217 **"The facts we turn up":** Ibid.

217 **a group of officers surrounded:** Silkie Carlo's recollection of event in interview with the author, 2022.

217 **the Met's official report:** "MPS LFR Deployments 2020–Date."

217 **"ongoing work to tackle":** "Four Arrests in Live Facial Recognition Operation in Central London," Borough of Hounslow *Herald,* January 29, 2022, https://hounslowherald.com/four-arrests-in-live-facial-recognition-operation-in-central-london-p15361-249.htm.

218 **The Met would not say:** Author's email exchange with DCI Jamie Townsend, Met Ops 3, May 16, 2022.

218 **a biometric identifier embedded:** "e-Passports," U.S. Department of Homeland Security, https://www.dhs.gov/e-passports. The presence of a chip in a U.S. passport is indicated by a golden logo resembling a camera on the cover.

218 **candid human rights assessment:** Pete Fussey and Daragh Murray, *Independent Report on the London Metropolitan Police Service's Trial of Live Facial Recognition Technology,* University of Essex, July 2019, https://repository.essex.ac.uk/24946/1/London-Met-Police-Trial-of-Facial-Recognition-Tech-Report-2.pdf.

218 **Fussey and I met:** Author's interview with Pete Fussey, 2022.

219 **"large amount of information":** Author's email exchange with DCI Jamie Townsend, May 16, 2022.

219 **"pilot and evaluation stage":** Author's email exchange with Paul Fisher, South Wales Police news manager, April 20, 2022.

219 **that might be caught:** Around the same time that Fussey's report came out, a public affairs manager and father of two named Ed Bridges sued the South Wales Police after his face was scanned while he was out Christmas shopping and again while he was attending a political protest. In 2020, an appeals court ruled in Bridges's favor, saying that his human rights had been violated, but it also said that the use of facial recognition could be lawful, as long as police created local rules and policies for its use. So that's what they've done.

219 **"I think live facial recognition":** Author's interview with Pete Fussey, 2022.

219 **A high-end casino:** Author visited the casino and toured the security room in London, 2022.

219 **"white-glove service":** Author's interview with Ryan Best, 2022.

220 **"The primary use case":** Ibid.

220 **a security company called Facewatch:** "Facewatch's facial recognition for retail is proven to reduce theft in your business by at least 35% in the first year." "Facial Recognition for Retail Security," Facewatch.co.uk, https://www.facewatch.co.uk/facial-recognition-for-retail-sector.

220 **She had gone there:** Author's interview with Silkie Carlo, 2022.

221 **FindFace was essentially:** Shaun Walker, "Face Recognition App Taking Russia by Storm May Bring End to Public Anonymity," *The Guardian,* May 17, 2016.

221 **FindFace went viral:** Ethan Chiel, "This Face Recognition Company Is Causing Havoc in Russia—and Could Come to the U.S. Soon," Fusion, April 29, 2016. (Fusion is now Splinter News.)

221 **"fucking cool algorithm":** This is a translation of a message originally in Russian: Andrey Mima, VK, March 24, 2016, https://vk.com/wall66559_67051.

221 **A few weeks later:** Kevin Rothrock, "The Russian Art of Meta-Stalking," Global Voices, April 7, 2016.

221 **The contrast was startling:** Marta Casal, " 'Your Face Is Big Data,' the Danger of Facial Recognition," Medium, October 3, 2019.

221 **"Your Face Is Big Data":** Egor Tsvetkov, "Your Face Is Big Data," https://cargocollective.com/egortsvetkov/Your-Face-Is-Big-Data.

221 **FindFace took a darker turn:** Kevin Rothrock, "Facial Recognition Service Becomes a Weapon Against Russian Porn Actresses," Global Voices, April 22, 2016.

221 **to identify two arsonists:** "The End of Privacy: 'Meduza' Takes a Hard Look at FindFace and the Looming Prospect of Total Surveillance," Meduza, July 14, 2016.

222 **Based on images:** " 'We'll Find You, Too,' " Meduza, July 8, 2017.

222 **NtechLab shut FindFace down:** Author's email exchange with Nikolay Grunin, head of PR for NtechLab, October 9, 2019. "In 2018, our company obtained the rights to the 'findface' brandname and shut the service down," wrote Mr. Grunin.

222 **deciding to sell:** Author's interview with Alexander Tomas, communications director for NtechLab, 2021.

222 **a "safe city" initiative:** Thomas Brewster, "Remember FindFace? The Russian Facial Recognition Company Just Turned On a Massive, Multimillion-Dollar Moscow Surveillance System," *Forbes,* January 29, 2020.

222 **She had an unpleasant:** Author's interview with Anna Kuznetsova, with translator Slava Malamud, 2022.

222 **She decided to give it a try:** Ibid.

222 **a black market:** Andrey Kaganskikh, "Big Brother Wholesale and Retail," YouTube, December 5, 2019, cited in "As Moscow's Facial Recognition System Activates, Journalists Find Access to It for Sale on the Black Market," Meduza, December 5, 2019.

223 **The report contained:** Kuznetsova shared parts of the report. On file with author.

223 **Roskomsvoboda publicized:** Umberto Bacchi, "Face for Sale: Leaks and Lawsuits Blight Russia Facial Recognition," Reuters, November 9, 2020.

223 **There were errors:** Author's interviews with Ekaterina Ryzhova, lawyer with Roskomsvoboda, 2020–2021.

223 **who would be stopped:** "История Сергея Межуева: Первый Кейс По

Ошибке Системы Распознавания Лиц в МетроV ("The Story of Sergei Mezhuev: The First Case of a Mistake in the Face Recognition System in the Subway"), Roskomsvoboda, November 19, 2020.

223 **By 2021, the system:** "«Происходящее На Видео Противоречит Содержанию Протокола». Юрист Екатерина Рыжова о Распознавании Протестующих По Камерам" ("'What Is Happening on the Video Contradicts the Content of the Protocol': Lawyer Ekaterina Ryzhova on Recognizing Protesters by Cameras"), OVD-Info, January 17, 2022.

223 **"Every society needs":** Author's interview with Anna Kuznetsova, 2022.

224 **He loved the chaotic energy:** Author's interview with Paul Mozur, 2022.

224 **"I lived in a dorm":** Ibid.

224 **to thwart tissue paper thieves:** "Beijing Park Dispenses Loo Roll Using Facial Recognition," BBC News, March 20, 2017.

224 **"like a bazooka":** Author's interview with Paul Mozur, 2022.

225 **In the seaside town of Xiamen:** Paul Mozur and Aaron Krolik, "A Surveillance Net Blankets China's Cities, Giving Police Vast Powers," *New York Times,* December 18, 2019.

225 **"Some of the use cases":** Author's interview with Paul Mozur, 2022.

225 **In Suzhou:** Amy Qin, "Chinese City Uses Facial Recognition to Shame Pajama Wearers," *New York Times,* January 21, 2020.

225 **In Ningbo:** "Chinese AI Caught Out by Face in Bus Ad," BBC News, November 27, 2018.

225 **Only VIPs could:** Paul Mozur, Muyi Xiao, and John Liu, "How China Is Policing the Future," *New York Times,* June 26, 2022.

225 **"minority identification":** Paul Mozur, "One Month, 500,000 Face Scans: How China Is Using A.I. to Profile a Minority," *New York Times,* April 14, 2019.

225 **China had monitored:** BuzzFeed News investigation by Megha Rajagopalan, Alison Killing, and Christo Buschek. Five-part series starting with "China Built a Vast New Infrastructure to Imprison Uighurs," BuzzFeed News, August 27, 2020.

226 **"Tech became so linked":** Author's interview with Paul Mozur, 2022.

227 **"We freaked out":** Ibid.

227 **the police said:** Paul Mozur's recollection of the encounter in interview with the author, 2022.

227 **In 2021:** Julian Barnes and Adam Goldman, "C.I.A. Admits to Losing Informants," *New York Times,* October 5, 2021.

Chapter 23: A Rickety Surveillance State

228 **"We will never give up":** Brian Naylor, "Read Trump's Jan. 6 Speech, a Key Part of Impeachment Trial," NPR, February 10, 2021.

229 **Clearview saw a surge:** Kashmir Hill, "The Facial-Recognition App Clearview Sees a Spike in Use After Capitol Attack," *New York Times,* January 9, 2021.

229 **"tragic and appalling":** Author's interview with Hoan Ton-That for *The New York Times,* 2021.
229 **"In 2017, I was like":** Ibid.
229 **"You see a lot of detractors":** Ibid.
230 **It was around that time:** Author's interview with Armando Aguilar, 2022.
231 **Miami detectives have since used:** Ibid.
231 **"People are right":** Ibid.
232 **"This generation posts everything":** One of the Real Time Crime Center detectives during author's visit in March 2022.
233 **"The system becomes":** Alejandro Gutierrez during author's visit to RTCC, March 2022.
233 **"Let's keep the camera":** An RTCC detective during author's visit to RTCC, March 2022.
233 **"It's like playing":** An RTCC detective during author's visit to RTCC, March 2022.
234 **"Vendors always make":** Alejandro Gutierrez during author's visit to RTCC, March 2022.
234 **"Thanks to all the technology":** Armando Aguilar during author's visit to RTCC, March 2022.
235 **gave the city grants:** Interview with Alejandro Gutierrez during author's visit to RTCC, March 2022.
235 **"You need to get out there":** Armando Aguilar during author's visit to RTCC, March 2022.

Chapter 24: Fighting Back

237 **for free to the Ukrainian government:** Drew Harwell, "Ukraine Is Scanning Faces of Dead Russians, Then Contacting the Mothers," *Washington Post,* April 15, 2022.
237 **When a defense attorney:** Kashmir Hill, "Clearview AI, Used by Police to Find Criminals, Is Now in Public Defenders' Hands," *New York Times,* September 18, 2022.
237 **But some in the public defense:** Hill, "Clearview AI, Used by Police to Find Criminals, Is Now in Public Defenders' Hands."
237 **"The hill I will die on":** Author's interview with Evan Greer, 2021.
238 **"Our message for Congress":** Fight for the Future, "We Scanned Thousands of Faces in DC Today to Show Why Facial Recognition Surveillance Should Be Banned," Medium, November 18, 2019.
238 **They have staved off:** Author's interview with Jay Stanley, 2021. Also Brian Hochman, *The Listeners: A History of Wiretapping in the United States* (Cambridge, Mass.: Harvard University Press, 2022).
239 **"It was somewhere between striking":** Author's interview with Brad Smith, 2022.
239 **"facial recognition technology":** Ed Markey, "Senators Markey & Merkley and Reps. Jayapal & Pressley Urge Federal Agencies to End Use of Clearview

AI Facial Recognition Technology," February 9, 2022, https://www.markey.senate.gov/news/press-releases/senators-markey-and-merkley-and-reps-jayapal_pressley-urge-federal-agencies-to-end-use-of-clearview-ai-facial-recognition-technology.

240 **"Recent NIST testing":** Hoan Ton-That, "Congressional Statement"; email sent from Lisa Linden to Kashmir Hill, February 9, 2022.

240 **ranked among the world's most accurate:** Kashmir Hill, "Clearview AI Finally Takes Part in a Federal Accuracy Test," *New York Times,* October 28, 2021; Kashmir Hill, "Clearview AI Does Well in Another Round of Facial Recognition Accuracy Tests," *New York Times,* November 23, 2021.

240 **NIST's "report card":** Datasheet, clearviewai_000, National Institute of Standards and Technology, https://pages.nist.gov/frvt/reportcards/1N/clearviewai_000.pdf; Developer Gains, clearviewai_000, National Institute of Standards and Technology, https://pages.nist.gov/frvt/reportcards/11/clearviewai_000.html.

241 **"It's companies you've never":** Author's interview with Alvaro Bedoya, 2022.

241 **"People into their privacy":** Scott Urban, quoted in Kashmir Hill, "Activate This 'Bracelet of Silence,' and Alexa Can't Eavesdrop," *New York Times,* February 14, 2020.

241 **he invented CV Dazzle:** Adam Harvey Studio, "CV Dazzle," https://adam.harvey.studio/cvdazzle/.

241 **"privacy armor":** Law professor Elizabeth Joh uses another term, "privacy protests." Elizabeth Joh, "Privacy Protests: Surveillance Evasion and Fourth Amendment Suspicion," *Arizona Law Review* 55, no. 4 (2013): 997–1029.

241 **despite his tweets of protest:** Author's interview with Adam Harvey, 2021.

242 **more than six thousand times:** Spreadsheet obtained by author through a public records request to the University of Washington.

242 **After Harvey documented:** Adam Harvey and Jules LaPlace, Exposing.ai, https://www.exposing.ai/datasets/.

242 **"I was interrupting":** Author's interview with Adam Harvey, 2021.

242 **"Our goal is to make Clearview":** Ben Zhao, quoted in Kashmir Hill, "This Tool Could Protect Your Photos from Facial Recognition," *New York Times,* August 3, 2020.

243 **During an all-hands meeting:** Ryan Mac, "Facebook Considers Facial Recognition for Smart Glasses," BuzzFeed News, February 25, 2021.

243 **"When AR glasses are common":** Recording of an all-hands meeting at Facebook provided by an anonymous source, February 25, 2021.

243 **"The Ugly":** Ryan Mac, Charlie Warzel, and Alex Kantrowitz, "Top Facebook Executive Defended Data Collection in 2016 Memo—and Warned That Facebook Could Get People Killed," BuzzFeed News, March 29, 2018.

244 **according to the company:** Exchange with Facebook's communications team during the fact-checking for this book, 2023.

244 **a fifty-page memo:** Elizabeth Dwoskin, "Inside Facebook's Decision to Eliminate Facial Recognition—for Now," *Washington Post,* November 5, 2021.

244 **At the end of 2021:** Kashmir Hill and Ryan Mac, "Facebook Plans to Shut Down Its Facial Recognition System," *New York Times,* November 2, 2021.
244 **Privacy advocates:** Author's interview with Adam Schwartz for *The New York Times,* 2021.
244 **"It's like putting the weapons":** Author's interview with a former Facebook product manager who spoke on condition of not being named, 2021.
244 **When I asked a Facebook spokesman:** Author's email exchange with Jason Grosse, 2021.

Chapter 25: Tech Issues

245 **"There's something wrong":** Author's interviews with Hoan Ton-That and Lisa Linden, 2021.
245 **"Life stopped":** Ibid.
246 **"I started finding more":** Ibid.
246 **"It's a time machine":** Ibid.
247 **a recent report:** "Facial Recognition Technology: Current and Planned Uses by Federal Agencies," U.S. Government Accountability Office, August 24, 2021, https://www.gao.gov/assets/gao-21-526.pdf.
248 **"That's probably not good":** Author's interview with Charles Johnson, 2021.
248 **signed a legal document:** Signed by Richard Schwartz, Hoan Ton-That, Hal Lambert and James Lang, "Action by Unanimous Written Consent of the Board of Directors of Clearview AI," May 21, 2021, copy provided to author by Charles Johnson. Signed by Hoan Ton-That, "Clearview AI, Inc. Buy-back Notice," October 9, 2021, copy provided to author by Charles Johnson.
248 **According to Johnson:** Ibid.
248 **Johnson's true goal:** Author's interviews with Charles Johnson, 2022.
249 **said they'd signed an NDA:** Author's interview with Sam Waxman, 2022.
249 **soon sign a contract:** Kashmir Hill, "Air Force Taps Clearview AI to Research Face-Identifying A.R. Glasses," *New York Times,* February 4, 2022.
250 **"Hoan, stay focused":** Author's interviews with Hoan Ton-That and Lisa Linden, 2021.
251 **tried to convince other doctors:** Rebecca Davis, "The Doctor Who Championed Hand-Washing and Briefly Saved Lives," NPR, January 12, 2015.
251 **"It's time for the world":** Ibid.

INDEX

Abrams, Dan, 210
Abrams, Floyd, 205–207, 208–211, 212
ACLU (American Civil Liberties Union)
  Abrams case and, 205, 208–209, 211–213
  Facebook and, 143
  Ferg-Cadima and, 82–83, 85, 87
  opposition from, 62, 65, 202–204
  Project on Speech, Privacy, and Technology and, 202
  Rekognition and, 137–138, 162
  Williams and, 182
Acquisti, Alessandro, 106, 107–108, 109, 110, 126, 143
Afghanistan, war in, 154
age
  accuracy of detecting, 124
  differences in facial recognition and, 70
Aguilar, Armando, 230–231, 233, 234, 235
Aibo, 151
Airbnb
  Alströmer and, 10
  sex workers and, 199
  Ton-That and, 5
Ake, David, 269n42
Alamy, 192
algorithmic bias, 178
Ali, Alaa, 171–174
All Things Digital, 102
Alphabet, 151
al-Qaeda, 64
Alströmer, Gustaf, 10
alt-right
  Smartcheckr and, 93, 94
  Ton-That and, 9
Amazon
  Floyd and, 239
  Rekognition and, 116, 137–138, 162, 238
Amendments
  First, 15, 206–207, 208–209, 212–213, 306n206
  Fourth, 141
American Civil Liberties Union (ACLU)
  Abrams case and, 205, 208–209, 211–213
  Facebook and, 143
  Ferg-Cadima and, 82–83, 85, 87
  opposition from, 62, 65, 202–204
  Project on Speech, Privacy, and Technology and, 202
  Rekognition and, 137–138, 162
  Williams and, 182
American Sociological Society, 23
Ancestry, 248
Android
  tracking and, 141
  unlocking and, 109, 142, 179
AngelList, 8, 81
Annan, Kofi, 104
anonymity, protections afforded with, 40
anticheating vigilantes, 33
antidiscrimination laws, 188
Antifascist Coalition, 54–55
antihacking laws, 117
anti-Semitism, 94

anti-wiretapping laws, 238
AOL, 150
Apple
  Clearview AI and, 165
  Franken and, 141
  iPhone and, 6, 9
  Leyvand and, 152
  phone locks and, ix
  Polar Rose and, 108
Arbitron, 46, 109, 270–271n44
Aristotle, 17
Armey, Dick, 62, 63–64, 65–66
Art Selfie app, 151
artificial intelligence, early efforts regarding, 36–37
arXiv.org, 72–73
Ashley Madison, 33, 201
Aspire Mirror, 156
Assange, Julian, 106
AT&T, 150
atavism, 23
Atick, Joseph, 63–64, 65, 66, 68–69, 71
Atkinson, Benjamin, 301n175
Atlanta Police Department, 157–159, 160
atomic bomb, 48
augmented reality, 164, 243, 244, 249
Australian Federal Police, 136
automated pattern recognition, early work in, 37
automatic target recognition, 43
Autonomous Land Vehicle ("Alvin"), 42, 43, 44
"average face," 48

background photo searches, 247
backpropagation, 147
Baidu, neural networks technology and, 74
"Ban Facial Recognition" campaign, 238
Bannon, Steve, 56
bans on face surveillance, 138
bathrooms, 224
Bayesian image processing, 104
Bazile, Pawl, 272n50
Beagle, HMS, 18
Bear Stearns, 111
Bedoya, Alvaro
  congressional work of, 140–144, 146, 148–151
  at FTC, 240–241
  at Georgetown, 151, 156
Been Verified, 58
Bell Labs, 37
Bentham, Jeremy, 226
Bernstein, Daniel, 306n206
Bernstein v. Department of Justice, 306n206
Bertillon, Alphonse, 21–22, 38
Besson, Greg, 128–129, 132
Betaworks, 112
Bezos, Jeff, 115–116
bias
  algorithmic, 178
  confirmation, 181
  racial, 156, 178–179
  testing and, 178, 179–180, 240
Biden, Jill, 280–281n79
Biden, Joe, 228, 280–281n79
Big Brother Watch, 215
Biometric Information Privacy Act (BIPA; Illinois)
  Clearview AI and, 158, 204, 205, 206, 213
  lawsuits filed under, 151–152
  passage of, 86
  SceneTap and, 122
biometrics. *See also* facial recognition; fingerprints/fingerprinting
  Abrams case and, 209, 211, 212
  care with in Illinois, 151, 158
  communities of color and, 239
  lack of laws regarding, 100
  legislation protecting, 82–87, 223, 240
  orders to delete information on, 192
  possible expansion of use of, 194
  use of in United Kingdom, 215–220
BIPA (Biometric Information Privacy Act; Illinois)
  Clearview AI and, 158, 204, 205, 206, 213
  lawsuits filed under, 151–152
  passage of, 86
  SceneTap and, 122
Bitcoin, 81, 222–223

Black Hat, 108, 110
Black Lives Matter, 11
Blackwater, 51–52
Bledsoe, Woody, 37, 38, 39, 40, 125
Bloomberg LP, 81
Blue Lives Matter, 232
BMI predictors, 32
Booker, Cory, 11
border walls, sensor-based, 57
Borthwick, John, 112–113
Boston Marathon bombing, 287n106
Bosworth, Andrew "Boz," 243–244
Brandeis, Louis D., viii, 204–205
Bratton, William, 129–130
Bridges, Ed, 308n219
Brill, Julie, 122, 125
Broderick, Matthew, 117
"broken windows" policing, 129
Brown, Dan, 247
Brown, Michael, 11
Brululo, Bridget, 292n134
Buckby, Jack, 272n50
Buolamwini, Joy, 156
Burning Man, 8–9, 10
Bush, George W., ix, 11, 64, 157, 209
Buzz, 102
BuzzFeed News, 165–166

Cagle, Matt, 137–138
Cahill Gordon & Reindel, 205–206
Cambridge Analytica scandal, 6, 92
cameras. *See also* photographs/photography
  Clearview AI Camera, 111
  Eastman Kodak film camera, viii
  Insight Camera, 187
  skin tone and, 179
  speed cameras, 238
  True Depth camera, 109
Capitol insurrection, 228–230
Carlo, Silkie, 215–217, 218, 220
Carlson, Tucker, 11
Cato Unbound, 261n15
Catsimatidis, Andrea, 114
Catsimatidis, John, 114
cease-and-desist letters, 165
census bureaus, 24–25
Central Intelligence Agency (CIA), 38, 125, 227, 268–269n38
Ceph, 79
Cernovich, Mike, 12, 53
Chaos Computer Club, 190
ChatGPT, 73
Chau, Ed, 186
Chicago Police Department, 155
child crimes investigators, 134–136
child sexual abuse material (CSAM), 135. *See also* pornography
China
  scoring of citizens by, 34
  surveillance in, 224–227
Churchill, Winston, 41
CIA (Central Intelligence Agency), 38, 125, 227, 268–269n38
Citizens United, 208, 306n206
Clarifai, 32–33
Clarium Capital, 14
ClassPass, 8
Clearview AI
  absence of headquarters for, x–xi, xii
  access requests and, 190–192
  ACLU and, 202, 203–204, 205
  AR glasses and, 249–250
  attempts to shut down communication by, xv, xvii–xviii
  author's contact with representatives of, 160–166
  bans on, 165
  capabilities of, ix–x
  Capitol insurrection and, 229–230
  concerns regarding weaponization and, xv–xvii
  defense of, 157–159
  effectiveness of, xii
  emergence of, 94
  fines assessed to, 193–194, 230
  growth of database for, 246–247
  hit rate of, 133
  international backlash against, 192–193
  international expansion and, 137
  investment efforts and, 111–120, 136. *see also individual investors*
  Johnson and, 95–96
  law enforcement and, 128, 130–139
  lawsuits against, 165, 204–206, 209–213, 248

Clearview AI (*cont'd*):
  Leone and, 113–114
  NYT article on, 164–165, 187, 190, 194, 204, 230
  opposition to, 237–238
  pandemic and, 186–187
  police reactions to, xii–xiv
  potential legal challenges to, 117
  push for ethical use and, 239
  results blocked by, xvii–xviii, 162–163
  Scalzo and, 111–112, 113, 188–189
  third-party testing of, 240
  tip regarding, vii–viii, ix
  Ukraine invasion and, 237
  wrongful arrests and, 183
Clearview AI Camera, 111
Clearview AI Check-In, 111–112
Clearview AI Search, 111
Clement, Paul
  attempts to contact, 160
  on Clearview AI's capabilities, ix–x
  lack of response from, xi
  legal memos from, 134, 157–158
Clinton, Bill, 209, 259n10
Clinton, Chelsea, 9–10, 259n10
Clinton, Hillary
  "deplorables" comment by, 50–51
  election loss of, 88
  facial recognition and, 104
  false claims about, 94
  Trump and, 16, 52
Cohen, Chuck, 133
Colatosti, Tom, 62, 65, 66, 71
Comet Ping Pong, 55
Computer Fraud and Abuse Act (1986), 117–118
computers
  early, 36
  reliance on, 36
confirmation bias, 181
Constitution
  First Amendment, 15, 206–207, 208–209, 212–213, 306n206
  Fourth Amendment, 141
consumer protection laws, 205
contact tracing, 186
Coolidge, Calvin, 11
Couchsurfing, 81
Coulson, Jennifer, 180–181
Coulter, Ann, 119
Covid-19, 185–187, 209, 214
Crime and the Man (Hooton), 25, 26
CrimeDex, xii, 134
criminal detectors, 31
"criminal face," 31
Criminal Justice Information Services Division, 299n158
Criminal Man (Lombroso), 22–23, 38
crisis communications, 161
Cruise, Tom, 123
Crunchbase, 79
Cruz, Ted, 11
cryptocurrency, 81
CSAM (child sexual abuse material), 135. *See also* pornography
CV Dazzle, 241

Dahua, 226
Daily Caller, 11
Daily News (New York), 28, 61
dancing, illegal, 129
Darling, Ela, 198–199
DARPA (Defense Advanced Research Projects Agency), 42–43, 67
Darwin, Charles, 18, 19–20
Darwin, Erasmus, 18, 19
data, labeled, 67
data and rate limiting, 58
data brokers, 204
Datacube, 46
David (pseudonym), 197–198, 199–200, 201
"Dawn of Robot Surveillance, The," 202
deep learning, 147
DeepFace, 147–148
Defense Advanced Research Projects Agency (DARPA), 42–43, 67
Defense Department, 66
Del Greco, Kimberly, 299n158
"demographic effects," 178
Department of Defense, 66
Department of Energy, 247
Department of Health and Human Services, 247
Department of Homeland Security, 134–136, 235

Department of Justice, 66, 247
Department of State, 66, 68–69
DeploraBall, 53–55, 56–57
"deplorables" comment, 50–51
Detroit Police Department, 169–177, 183
direct brain interface, 145
Division of Privacy and Identity Protection, 122
dlib, 77, 195
DNA collection, 226, 248
Dorsey, Jack, 95
Douglas, Heather, 48, 279n74
DPA (Hamburg Data Protection Authority), 192
driver's license photographs
  law enforcement access to, xiii, 143–144, 156, 177–178, 203, 234
  wrongful identification and, 69, 176, 180
drones, Smartcheckr and, 57–58
Duke MTMC, 242
Dumanski, Keith, 160–161
Dvach (2chan), 221–222

Eastman Kodak film camera, viii. *See also* Kodak
Edelson, 205
Eidelman, Vera
  Abrams and, 207
  Abrams case and, 204, 205, 208–209, 211–212
  with ACLU, 202
  background of, 203
eigenfaces, 46–47, 60, 63
Eisenhower, Dwight D., 42
Electronic Frontier Foundation, 306n206
Ellsberg, Daniel, 207
Ellwood, Charles, 23
embodied internet, 145–146
Energy Department, 247
Epstein, Jeffrey, 270n43
Ethereum, 81
eugenics, 20, 25–26
Everyone, 9
Expando, 6
Exposing.ai, 242
"Face Facts" workshop, 121–122, 123–127
Face Recognition Vendor Test, 99
Face Unlock, 109, 179
Facebook
  AR glasses and, 243, 244
  automatic tagging on, viii–ix, 142, 144
  BIPA and, 151–152, 204, 205
  Cambridge Analytica scandal and, 92
  Clearview AI and, vii, 96, 158, 165
  "Face Facts" workshop and, 121–122
  Franken and, 141, 143–144
  FTC and, 123
  identification technology and, 144–148
  Luckey and, 57
  monetization and, 6–7
  neural networks technology and, 74
  photo experiment and, 106–108, 126
  pornography and, 198–200
  public profile photos and, 142–143
  racial bias and, 178–179
  Saverin and, 94
  scraping and, 58, 117, 144, 195, 242
  Smartcheckr and, 53
  State Privacy and Security Coalition and, 150
  Thiel and, xi, 13
  third-party apps for, 4–6
  Ton-That and, 5
  turning off facial recognition, 244
  use of contact lists by, 7
Face.com, 143, 146
Facewatch, 220
facial recognition. *See also* Clearview AI; Rekognition; Smartcheckr
  after September 11, 64–65
  in casinos, 62
  early adopters of, 62–63
  early efforts regarding, 38–39, 41–42, 44, 45–49
  human abilities for, 45
  increased government interest in, 66–67
  Johnson and Ton-That's initial ideas for, 16

facial recognition (*cont'd*):
possible repercussions of, 39–40
roots of, 17–26
at Super Bowl, 60–62, 63
facial recognition units, mobile, 214–219
Facial Recognition Vendor Test, 68, 99, 178. *See also* NIST (National Institute of Standards and Technology)
Faraday bags, 227
FarmVille, 4–5, 6
"fatface," 32
Fawkes, 242
FBI (Federal Bureau of Investigation)
Capitol insurrection and, 229
Congress and, 142, 143–144, 158
Government Accountability Office and, 247
NIST and, 70, 71
NYPD and, 132
Federal Police of Brazil, 136
Federal Trade Commission (FTC), 121–123, 126–127, 143, 240
FERET, 67–68, 103
Ferg-Cadima, James, 82–87, 151, 158
Ferrara, Nicholas, xii–xiii
Fight for the Future, 238
Financial Crimes Task Force, 128–131
FindFace, 34–35, 220–222
Findley, Josh, 134–136
fingerprints/fingerprinting
early efforts regarding, 22
Pay by Touch and, 82, 84–87
theft of, 84
First 48, The, 230
First Amendment
Abrams case and, 206–207, 208–209, 212–213
Gawker and, 15
violence and, 306n206
FitMob, 8
Flanagan, Chris, 129, 131
Flickr, 199, 242, 246–247
Flipshot, 9
Floyd, George, 207, 239
FOIA (Freedom of Information Act), 141, 157
4Chan, 94
Fourth Amendment, 141
Franken, Al, 140–144, 148–151
Freedom of Information Act (FOIA), 141, 157
FreeOnes, 197
Frydman, Ken, 28
FTC (Federal Trade Commission), 121–123, 126–127, 143, 240
Fuller, Matt, 274n52
Fussey, Pete, 218–219, 235

Gaetz, Matt, 285n95
Galaxy Nexus, 109
Galton, Francis, 17–20, 22, 25, 33
GAO (General Accounting Office), 62, 65–66
Garrison, Jessica Medeiros, 138–139, 160
Garvie, Clare, 156–157, 178
Gaslight lounge, 50–51
Gawker Media
Johnson and, 11, 12
Pay by Touch and, 85
Thiel and, 14–15
Ton-That and, xi, 7, 93, 116, 164
gaydar, 31
GDPR (General Data Protection Regulation), 191
gender
Congressional representation and, 89
differences in facial recognition and, 48, 69–70, 124, 125, 156, 178, 240
IQ and, 32
General Accounting Office (GAO), 62, 65–66
General Data Protection Regulation (GDPR), 191
Genetic Information Nondiscrimination Act (GINA), 83
Gibson, William, 113
Giesea, Jeff, 53
Gilbert, John J., III, 129
GINA (Genetic Information Nondiscrimination Act), 83
Girard, René, 12
GitHub, 72, 74

Giuliani, Rudy
Schwartz and, 27–28, 29, 89, 90, 129, 161
Waxman and, 80
Global Positioning System (GPS), 43
Gmail, 102
Gone Wild, 200
Good, John, x, 160
Google
AI Principles of, 108
BIPA and, 151
Clearview AI and, 96, 165
CSAM and, 135
diverse datasets and, 179
"Face Facts" workshop and, 121–122
Face Unlock and, 109, 179
Franken and, 141, 142
FTC and, 123
Goggles and, 99–100, 102
hesitation of regarding facial recognition technology, 99–110
identification technology and, 145
Images, 79
lawsuits against, 306n206
Lunar XPRIZE and, 192
Maps and, 100–101, 102
mining of Gmail by, 102
monetization and, 6–7
neural networks technology and, 74
phone locks and, ix
Photos app, 272n48
PittPatt and, 108–109, 110
Schmidt and, 27
State Privacy and Security Coalition and, 150
Street View and, 100–101, 102
TensorFlow and, 208
ViddyHo and, 7
GotNews, 11–12
Government Accountability Office, 247
GPS (Global Positioning System), 43
Greenwald, Glenn, 149
Greer, Evan, 237–238
Grewal, Gurbir, 165
Gristedes, 114
Grossenbacher, Timo, 194–196
Grother, Patrick, 69–70
Grunin, Nikolay, 309n222
Guardian, The, 149
gunshot detection systems, 232–233, 234
Gutierrez, Alejandro ("Gooty"), 231–232, 234

Hacker News, 4
Hamburg Data Protection Authority (DPA), 192
Haralick, Robert, 271n47
Harrelson, Woody, 156–157
Harris, Andy, 274n52
Harvard Law Review, on privacy, viii
Harvey, Adam, 241
Haskell, 4
Health and Human Services Department, 247
Hereditary Genius (Galton), 20
Hikvision, 177, 215–216, 226
Hinton, Geoffrey, 73, 74, 295n146
History of Animals (Aristotle), 17
Hogan, Hulk, 15
Hola, 275n58
Hollerith, Herman, 24–25
Homeland Security, Department of, 134–136, 235
Hooton, Earnest A., 25, 26
Hot Ones, 115
Howard, Zach, 145
Howell, Christopher, 207–208
HTML (Hyper Text Markup Language), 78
Huawei, 226
HuffPost, 165
Hungary, 56–59
Hyper Text Markup Language (HTML), 78

"I Have a Dream" speech (King), 39
IARPA (Intelligence Advanced Research Projects Activity), 106
IBM, 25, 156, 239
Incredibles, The, 119
Independent, The, 34
Indiana State Police, 133
Inmar Intelligence, 87
Insight Camera, 187

Instagram
  Brown and, 11
  Clearview AI and, vii
  Marx and, 192
  scraping and, 195
  Williams and, 182
Instaloader, 194–195
Instant Checkmate, 58
Intel, ix, 123–125
Intelligence Advanced Research Projects Activity (IARPA), 106
internet
  DARPA and, 43
  embodied, 145–146
  Trump and, 54
Internet Archive, 78
Interpol, 136
Intimcity, 221
investigative lead reports, 71, 131, 176, 180
iPhone
  developer tools for, 9
  release of, ix, 6
  unlocking and, 109
IQ, 32
"I've Just Seen a Face," 125
Iveco vans, 214

James, LeBron, 14
January 6 insurrection, 228–230
Java, 4
Jayapal, Pramila, 239
Je Suis Maidan, 222
Jenner, Kendall, 115
Jewel-Osco, 82
Jobs, Steve, ix
Johnson, Charles Carlisle "Chuck"
  background of, 11–12
  on blocking author's face, 163
  contact with, 165
  early plans of, 31, 34
  FindFace and, 220
  Gawker and, 15
  investors and, 116
  Jukic and, 134
  Orbán and, 56
  ouster of, 94–96
  at Republican Convention, 11, 15–16
  Scalzo and, 118–119
  Schwartz and, 29, 161–162
  Smartcheckr and, 52–53, 72, 79, 80
  Ton-That and, 12–14, 27, 33, 161–162, 247–249
  Trump and, 50, 51–52, 53, 54
Joint Terrorism Task Force, 132
Jukic, Marko, xvii–xviii, 134, 162
Justice Department, 66, 247

Kalanick, Travis, 189
Kanade, Takeo, 40, 41–42, 103
Keeper, 117
Kennedy, John F., 41
King, Martin Luther, Jr., 39–40
King-Hurley Research Group, 268–269n38
Kirenaga Partners, xiv–xv, 111, 112, 160
Knox, Belle, 198
Kodak, viii, 104, 179
Krizhevsky, Alex, 295n146
Kroger's, 87
Krolik, Aaron, 164
Kutcher, Ashton, 114–115
Kuznetsova, Anna, 222–223

L-1 Identity Solutions, 71
labeled data, 67
Lambert, Hal, 118
law enforcement. *See also* New York City Police Department (NYPD); *individual government entities*
  abuses by, 156–157
  Atlanta Police Department, 157–159, 160
  "broken windows" policing, 129
  Chicago Police Department, 155
  Clearview AI and, 128, 130–139
  Detroit Police Department, 169–177, 183
  Federal Police of Brazil, 136
  Franken and, 143–144
  Metropolitan Police (London), 214–219
  Miami Police Department, 230–236
  Michigan State Police, 180
  New York City Police Department (NYPD), 128–133, 138

- Ontario regional police services, 136
- PittPatt and, 105–106
- private surveillance and, 204
- Queensland Police Service, 136
- Royal Canadian Mounted Police, 136
- Schwartz and, 132–133
- South Wales Police, 219
- surveillance by, 155
- Swedish Police Authority, 136
- Toronto Police, 136

laws/legislation
- antidiscrimination laws, 188
- antihacking laws, 117
- anti-wiretapping laws, 238
- Biometric Information Privacy Act (BIPA; Illinois), 86, 122, 151–152, 158, 204, 205, 206, 213
- Computer Fraud and Abuse Act, 117–118
- consumer protection laws, 205
- Freedom of Information Act (FOIA), 141, 157
- General Data Protection Regulation (GDPR), 191
- Genetic Information Nondiscrimination Act (GINA), 83
- location privacy bill, 149–150
- privacy laws, 191
- Texas Capture or Use of Biometric Identifier Act, 83
- USA PATRIOT Act, 64

Lawyer Appreciation Day, 213
LeCun, Yann, 73
Leone, Doug, 113–114
Les Ambassadeurs, 219–220
Lewis, John, 138
Leyvand, Tommer, 144–146, 152
Linden, Lisa
- Abrams case and, 209
- Lawyer Appreciation Day and, 213
- NYT article and, 161
- Ton-That interviews and, 164, 245, 249, 250

LinkedIn
- Clearview AI and, vii, 112, 160, 165
- Clearview AI's lack of presence on, x
- Schwartz and, 94
- scraping and, 58, 117–118
- Smartcheckr and, 53
- use of contact lists by, 7

Lipton, Beryl, 157
Lisp, 4
Liu, Terence Z.
- background of, 75
- Clearview AI and, 240
- dlib and, 195
- Smartcheckr and, 75–77, 79, 81

location metadata, 206
location privacy, 149–150, 203
Lockheed Martin, 42
logic-based AI, 37
Lombroso, Cesare, 22–23, 25, 31, 38
Los Angeles Times, 61
Lost Symbol, The, 247
LOVEINT, 131
Luckey, Palmer, 57, 79
Lynch, Holly, 88, 89–93

MacAuley, Lacy, 54–56
machine learning, 72
machine reading, early work in, 37
Mackey, Douglass, 91, 92, 93–94
Madison Square Garden, 187–188
MAGA ETF, 118
"Magnifying Glass" exhibit, 40–41
Malchow, Joseph, 118
Manhattan Institute, 161
Manhattan Project, 279n74
Mao Zedong, 14
Maple, Jack, 129–130
Markey, Ed, 138, 165, 186
Martin Marietta, 42
Martinez, Freddy, 153–155, 157–159
Marx, Matthias, 190–192
McCartney, Paul, 125
McConnell, Mitch, 306n206
McInnes, Gavin, 50–51, 55
McKesson, DeRay, 11
Meadows, Mark, 156
Meetup, 81
MegaFace, 242
Megvii, 226, 242
Megyesy, Eugene, 56, 275n59
memetic desire, 12
Meta, 146. *See also* Facebook

metadata, location, 206
metaverse, 146
#MeToo movement, 151
Metropolitan Police (London), 214–219
Miami Police Department, 230–236
Michigan State Police, 180
Microsoft
  digital billboards and, ix
  neural networks technology and, 74
  PhotoDNA and, 135
  push for ethical use and, 239
  racial bias and, 156
  Smartcheckr and, 76
  Xbox and, 142
Minority Report, vii, 123, 125
Minsky, Marvin, 37, 43–44, 73
Miranda rights, 175
MIT Media Lab
  Buolamwini and, 156
  "Ricky Vaughn" Twitter account and, 94
  Turk and, 43–44, 46, 48, 60
Mithal, Maneesha, 121, 126, 143
mobile apps, 6
mobile facial recognition units, 214–219
Monroe, Marilyn, 41
More, Trenchard, 267n36
Moses, Robert, 27
Mossberg, Walt, 102–103
Mozur, Paul, 224–225, 226, 227
MuckRock, 157
mug shots
  ACLU and, 203
  Clearview AI and, 240
  Congressional hearings and, 143
  "criminal face" and, 31
  early facial recognition efforts and, vii–viii
  FBI and, 70, 71
  Florida's use of, 63
  Michigan's database including, 177–178
  NYPD and, 130, 131
  Rekognition and, 138
  scraping and, 33
  standardization of, 22
Myspace, 123
NAACP, 39
Nadler, Jerry, 89, 91–92
Narrative of an Explorer in Tropical South Africa (Galton), 18
National Crime Agency, 136
National Institute of Standards and Technology (NIST)
  bias problems and, 178, 179–180, 240
  Clearview AI and, 162
  Facial Recognition Vendor Test by, 68, 99, 178
  FERET and, 67–68, 103
  founding of, 67
  testing by, 67–70
National Press Club, 53, 55
National Security Agency (NSA), 149
NATO summit, 154
Nazi regime, 20, 25–26
NBC News, 94
NEC (Nippon Electric Company), 40–41, 180, 216, 240
Nehlen, Paul, 93
Nest, 151
Netflix, 73
neural networks technology
  DeepFace and, 147
  development of, 73–75
  Minsky and, 37, 73
  poor training and, 178
  power of, 203
  Schneiderman and, 103–104
  Smartcheckr and, 76
  SuperVision and, 147
Neven, Hartmut, 99, 101–102
Neven Vision, 99–100
New York City Police Department (NYPD), 138
New York Times Magazine, 209, 248
New York Times, The
  Abrams and, 206–207
  on Clearview AI, 164–165, 187, 190, 204, 230
  on Facebook, 107
  Super Bowl (2001) and, 62
  on supercomputers, 75
Newham Borough, 66
Nielsen, 44, 271n46
9/11, 28, 64, 235

Nippon Electric Company (NEC), 40–41, 180, 216, 240
NIST (National Institute of Standards and Technology)
  bias problems and, 178, 179–180, 240
  Clearview AI and, 162
  Facial Recognition Vendor Test by, 68, 99, 178
  FERET and, 67–68, 103
  founding of, 67
  testing by, 67–70
Nixon, Richard, 206–207
Northrop Grumman, 242
NSA (National Security Agency), 149
NtechLab, 220–221, 222, 240
Nvidia, 147
NYPD (New York City Police Department), 128–133, 138

Obama, Barack, 11, 122, 209
Ocasio-Cortez, Alexandria, 283n89
Occupy Chicago movement, 153–155
Occupy Wall Street movement, 153
Oculus Rift, 79
Oculus Virtual Reality, 57
Office of Personnel Management, 84
O'Harrow, Robert, 277n66
OkCupid, xvii, 33
Oliver, Michael, 183
Olympics, 62
On the Origin of Species (Darwin), 19
online ghosts, xiv
OnlyFans, 201
Ontario regional police services, 136
Oosto, 219
Open Society Foundations, 57
Open the Government, 155
OpenFace, 74
Operation Open Door, 165
Oppenheimer, J. Robert, 48
Opportunity America, 28
opposition research, extreme, 90–91
Orbán, Viktor, 56–57
Orlando, Florida, 137–138

Palantir, xi
pandemic, 185–187, 209, 214
Panopticon, 226
Panoramic Research Incorporated, 37–38
parabiosis, 261n15
Paul, Rand, 118
Paul Hastings, 80
Pay by Touch, 82, 84–87
PayPal, xi, 13, 77, 210–211
Pentagon, September 11 and, 64
Pentagon Papers, 206–207
Pentland, Alex "Sandy," 44
People You May Know feature, 198
Perceptrons (Minsky, Papert, et al.), 37
"Perpetual Line-up, The," 156
Persistent Surveillance Systems, 204
Phillips, P. Jonathon, 67–68, 70–71
PhotoDNA, 135
photographs/photography. *See also* cameras; driver's license photographs; mug shots
  background photo searches, 247
  physics of, 179
  public profile photos, 142–143
  punch photographs, 24–25
  stock photo sites, 192
  visa application photos, 68–69
PHP, 5
phrenology, 18
physiognomy, 13, 17–18, 19–20, 31, 40–41
Picasa, 101
PimEyes, 197–198, 199, 200–201
Pinterest, Smartcheckr and, 53
PitchBook, xi, xiv
Pittsburgh Pattern Recognition ("PittPatt")
  Acquisti and, 107, 108, 110, 126
  early work of, 105
  founding of, 104–105
  Google's purchase of, 103, 108–109, 110
  IARPA and, 106
  Intel and, 124
#PizzaGate conspiracy theory, 12, 55
Podesta, John, 259n10
Polar Rose, 108
pornography, 197–200, 221–222. *See also* child sexual abuse material (CSAM)
Pressley, Ayanna, 239

Prince, Erik, 51–52
principal components analysis, 47
privacy
  description of, viii
  laws regarding, 191
  right to, viii, 204–205
privacy paradox, 106, 107
privacy policies, 123
private right of action, 85, 86
private surveillance, law enforcement and, 204
probable cause, absence of, 180
probe images, 180
Project on Speech, Privacy, and Technology, 202
Project Veritas, 55
Proud Boys, 50
proxy networks, 58
Public by Default, 280n77
punch photographs, 24–25
Python, 194–195

Queensland Police Service, 136

race
  variation in accuracy and, 65, 70–71, 156, 178–179, 239, 240
  wrongful arrests and, 183
racism, 19, 20, 23, 33, 94, 259n10
railroad industry, 23–25
Railway News, The, 24
Raji, Inioluwa Deborah, 278n71
Rank One Computing, 180
RapidSOS, 232
Ravikant, Naval, 4–5, 6, 8, 118
Raviv, Shaun, 268n38
Raytheon, 60
Reagan, Ronald, 117
Real Time Crime Center (RTCC), 132, 230, 231–236
RealWear, 164
"Reclaim Your Face" campaign, 238
Reddit, 10, 53, 200
Reflectacles, 241
Reid, Randal, 183
Rekognition, 116, 137–138, 162, 238
repeat offenders
  eugenics and, 25
  identifying, 21–22
Republican Attorneys General Association, 138–139
Republican Party convention, 3, 10–11, 12, 13–14, 27
revenge porn, 200
"Ricky Vaughn" Twitter account, 93–94
"Right Wing Gossip Squad," 94
Rockefeller Foundation, 36
Roe, Phil, 274n52
Rogers, John, 84, 85
Rolling Stone, 11
Ronan, James, 301n175
Roskomsvoboda, 222–223
Royal Canadian Mounted Police, 136
Royal Geographical Society, 19
RTCC (Real Time Crime Center), 132, 230, 231–236
Rudin Management, 129
Russia, 34–35, 220–223, 237. *See also* Soviet Union

Salem, Mohammed, 171–174
Sampras, Pete, 104
Saverin, Eduardo, 94
Scalia, Antonio, 306n206
Scalzo, David
  author's contact with, xv
  frustrations of, 188–189
  investment in Clearview from, 111–112, 113, 116–117, 118–120
ScanAmerica, 270–271n44
SceneTap, 122
Schemm, Martin, 304n192
Schmidt, Eric, 27, 102–103, 145
Schneiderman, Henry, 103–106, 108–109
Schneier, Bruce, 61, 65
Schwartz, Richard
  Abrams and, 205–206
  attempts to contact, 160, 213
  background of, 27–29
  Besson and, 128, 129
  early plans of, 31, 33–34
  FindFace and, 220
  investors and, 116
  Johnson and, 95, 247–248
  law enforcement and, 132–133

Linden and, 161
Lynch and, 88, 89–90, 91, 92, 93
Montana and, 114
NYPD and, 129, 130, 131
opposition and, 138
Orbán and, 56
sale to businesses and, 187
Scalzo and, 111, 113, 118, 188–189
Smartcheckr and, 52, 72, 79, 89, 94
Ton-That and, 29–30
Waxman and, 80
Science, Policy, and the Value-Free Ideal (Douglas), 279n74
scraping
Abrams case and, 206, 208–209
Bedoya and, 144, 241
Borthwick and, 113
description of, 58
efforts to block, 58, 117–118, 165
Fawkes and, 242
Grossenbacher and, 194–195
LinkedIn and, 117–118, 165
Smartcheckr and, 92
Ton-That's use of, 78–79, 81, 116, 133, 162
Venmo and, 78–79
Vermont lawsuit and, 205
Search King, 306n206
Secret Service, 247
security theater, 65
Seeking, 201
Sejnowski, Terry, 268n37
self-driving cars, 42
selfies, as password alternative, 31
Semmelweis, Ignaz, 250–251
SenseTime, 226, 240, 242
September 11 attacks, 28, 64, 235
Sequoia Capital, 113
sex workers, 197–199, 221–222
Sherman, Rob, 144
Shirley cards, 179
ShopperTrak, 242
ShotSpotter, 232–233, 234
Showing Up for Racial Justice, 57
Shteyngart, Gary, 34
Six Flags amusement park, 151–152
Slack, 10
Slashdot, 104
Smartcheckr
contractors for, 81
DeploraBall and, 55, 56–57
development of, 73–79
founding of, 52–53
Hungary and, 55–56, 57–59
investment efforts and, 80, 89
lack of office for, 72
Lynch and, 90–93
MegaFace and, 242
name change for, 162
white nationalism and, 94
winding down of, 95, 249
smartphone economy, 6
smartphones
data brokers and, 204
Franken and, 141
iPhone, ix, 6, 9, 109
tracking and, 226
ubiquity of, 36
unlocking and, 109, 142
Smith, Brad, 239
Snap Map, 232
Snapchats, 232
Snowden, Edward, 149
social media, scraping and, 58, 92. *See also individual platforms*
Society of the Mind, The (Minsky), 270n44
Solove, Daniel, 126
Sony, 151
Soros, George, 57
South Wales Police, 219
Southern Poverty Law Center, 11, 50
Soviet Union, 42. *See also* Russia
speed cameras, 238
Spielberg, Steven, 123
Spitzer, Eliot, 161
Spotify, 8, 73
Sputnik, 42
Square, 5, 95
St. Petersburg Times, 61, 65
Stallone, Sylvester, 147–148
Standard & Poor's, 208
Star Trek, 105
State Department, 66, 68–69
State Privacy and Security Coalition, 150

sterilization, involuntary, 26. *See also* eugenics
stingrays, 155
stock photo sites, 192
"stop the steal," 228
Strategic Computing Initiative, 42
"Success Stories" marketing document, 137
Super Bowl (2001), 60–62, 63, 65
Super Sad True Love Story (Shteyngart), 34
supercomputers, 75–76
SuperVision, 146–147
Supreme Court
  Abrams and, 207
  Citizens United and, 306n206
  on First Amendment, 306n206
  involuntary sterilizations and, 26
  location privacy and, 203
  NAACP and, 39
  Pentagon Papers and, 207
Sutskever, Ilya, 295n146
Swedish Authority for Privacy Protection, 292n137
Swedish Police Authority, 136
Swisher, Kara, 102
symbolic AI, 37
Syndrome, 119

Taigman, Yaniv, 146–148
TechCrunch, 101
technical sweetness, 48, 148
Telegram, 222
Temple of Heaven, 224
TensorFlow, 208
Tesla, 73
Texas Capture or Use of Biometric Identifier Act, 83
Therrien, Daniel, 193
Thiel, Peter
  attempts to contact, 160
  background of, xi
  Clearview AI and, 111
  at DeploraBall, 55
  Gawker and, 14–15
  Kirenaga Partners and, xiv
  lawsuits against PayPal and, 210–211
  at Republican Convention, 15–16, 29
  Scalzo and, 119–120
  Smartcheckr and, 79–81, 89
  Ton-That and, 13, 14
  Trump administration and, 53
This Person Does Not Exist, 163, 195
Thorn, 115
TikTok, 224
Tinder, xvii, 34
toilet paper dispensers, 224
Tomas, Alexander, 309n222
Tompkins Square Park, 30
Tong, William, 139
Ton-That, Hoan
  Abrams and, 209–211
  AR glasses and, 249–250
  attempts to contact, 160
  attorneys general and, 138–139
  avoidance of past by, 247
  background of, 3–4
  Besson and, 128, 129
  Bezos and, 116
  bias and, 239
  on blocking author's face, 163
  Borthwick and, 113
  Capitol insurrection and, 229
  databases used by, 242
  dlib and, 195
  Dumanski and, 160–161
  early plans of, 31–35
  Facebook apps by, 5–6
  FindFace and, 220
  Gawker and, 15
  growth of database and, 246–247
  hit rate and, 133
  international backlash and, 193, 194
  investors and, 116
  Johnson and, 12–13, 27, 33, 94–96, 247–249
  Kutcher and, 115
  lawsuits and, 209–211, 213
  Leone and, 113–114
  Liu and, 75–76, 77
  Lynch and, 91, 92, 93
  meeting with, 161–164
  Montana and, 114
  in New York, 9
  NIST testing and, 240

NYPD and, 128, 129, 130, 131–132
opposition and, 138
Orbán and, 56
photos of, 8–9
politics of, 9–10
pseudonym use by, 134
at Republican Convention, 3, 10–11, 12, 13–14, 15–16
in San Francisco, 4–5
Scalzo and, 111, 112, 113, 118, 188–189
Schwartz and, 29–30
scraping and, 78–79, 81
self-assessment of, 250–251
Smartcheckr and, 52, 53, 58, 72–73, 74–75, 77–81, 89, 94
Thiel and, 79–80
Trump and, 50, 51–52
ViddyHo and, 7–8
Ton-That, Quynh-Du, 4
TopShot, 9
Toronto Police, 136
Touchfelt, 4
Traitwell, 248
Transportation Security Agency (TSA), 28
"Tremolo" (Ton-That), 8
True Depth camera, 109
true faces, 17
Trump, Donald
Cambridge Analytica scandal and, 92
as candidate, 3, 8, 16, 29
Capitol insurrection and, 228, 229
casinos owned by, 62
Congressional hearings and, 155–156
DeploraBall and, 53–54
election night and, 50–52
Johnson and, 95
Luckey and, 57
Lynch and, 88
Mackey and, 93–94
Orbán and, 56
Thiel and, xi
Ton-That's support of, 3, 10, 13, 16, 116
Trump, Melania, 14
Trump Hair, 9
TruthFinder, 58
TSA (Transportation Security Agency), 28
Tsvetkov, Egor, 221
Turk, Matthew
Autonomous Land Vehicle ("Alvin") and, 42
Facebook and, 152
at MIT Media Lab, 43–44, 45–47, 48–49, 60
timing of work of, 66
Turow, Joseph, 280n78
Tuskegee Institute, 279n74
23andMe, 248
Twitter
Cernovich and, 12
Clearview AI and, xvii, 165
FTC and, 123
Johnson and, 11, 95
machine learning and, 72
Mackey and, 93–94
Smartcheckr and, 53
Ton-That and, 5, 9
ViddyHo and, 7
Yiannopoulos and, 14
2chan, 221–222

Uber, 5, 189, 210
Ukraine, Russia's invasion of, 237
Ultra Music Festival, 231, 232, 233–234
undercover officers, dangers to, xvii
Urban, Scott, 241
U.S. Air Force, 249
USA PATRIOT Act, 64

Valleywag, 7–8, 85
Venmo, 77–79, 165
Veritas Strategic Partners Ltd., 90, 93, 150
Verizon, 149
VICE magazine, 50
Vicemo, 280n77
Victorian Era, 13, 17
ViddyHo, 7–8, 164
Vietnam War, 207
Vigilant Solutions, 204, 213
Viisage Technology, 60, 62–63
Viola, Andres Rafael, 136
Viola-Jones, 241

virtual private networks (VPN), 275n58
visa application photos, 68–69
Visionics Corporation, 63
VKontakte (VK), 34–35, 192, 220–221
Voyage of the Beagle, The (Darwin), 18
VPN (virtual private networks), 275n58
Vuzix, 164, 249

Wall Street Journal, The, 140–141, 186
Warhol, Andy, 13–14
Warren, Elizabeth, 94
Warren, Samuel D., Jr., viii
Washington Post, The, 93, 237
watch lists, 217, 219
Waxman, Samuel, 80, 95–96, 248–249
Wayback Machine, 78
weaponization concerns, xv–xvii
WeChat, 224
welfare "reform," 28
Wessler, Nathan, 202, 203, 205
WeWork, xi, 132, 161, 164
"What They Know" series, 140–141
white nationalism, 94. *See also* racism
WHO (World Health Organization), 185
Wikileaks, 106, 259n10
Williams, Melissa, 169–175, 182–183
Williams, Robert Julian-Borchak, 169–177, 180–184, 203
Williams, Serena, 156
Wolf, Helen Chan, 39
Wolosky, Lee, 209
World Exposition (Osaka), 40–41, 103, 180
World Health Organization (WHO), 185
World Naked Bike Ride, 200
World Trade Center, September 11 and, 28, 64
World Wide Web Wanderer, 275n58
wrongful arrests, 171–177, 180–184
Wyden, Ron, 165

Xbox, 142
Xiaomi, 224

Yahoo! 134–135, 150
Ybor City, 62–63, 65
Yglesias, Matt, 259n10
Yiannopoulos, Milo, 14
Yitu, 226
Yoo, John, 11
YOU Technology, 86
YouTube, 7, 99, 115

Zhao, Ben, 242
Zientek, Daniel, xiv
Zuckerberg, Mark
  aspirations of, 145–146
  Bosworth and, 243
  motto and, 152
  Taigman and, 146, 148
  Ton-That's comparison to, 94
Zuckerman, Mort, 28

ABOUT THE AUTHOR

KASHMIR HILL is a tech reporter at *The New York Times*. Her writing about the intersection of privacy and technology pioneered the genre. Hill has written for a number of publications, including *The New Yorker, The Washington Post, Gizmodo, Popular Science,* and *Forbes*.